Market Structure and Behavior

Market Structure and Behavior

Martin Shubik
with Richard Levitan

HARVARD UNIVERSITY PRESS
Cambridge, Massachusetts
and London, England 1980

Printed in the United States of America

Library of Congress Cataloging in Publication Data

Shubik, Martin.
Market structure and behavior.

Bibliography: p.
Includes index.
1. Industrial organization (Economic theory)
2. Oligopolies—Mathematical models. 3. Duopolies
—Mathematical models. 4. Game theory.
I. Levitan, Richard, 1927– joint author.
II. Title.
HD2326.S55 338.8'2'01 79-27108
ISBN 0-674-55026-9

To Oskar Morgenstern, friend and mentor

Preface

IN THIS BOOK WE EXPLORE a variety of oligopoly models based on an underlying structure of linear demand. All models noted are reproducible within the structure of a game. A later study will provide some tentative steps toward dynamics using an analytical and behavioral approach; theory and experimental results will be considered.

Chapter 1 sets the context for the approach blending classical oligopoly theory with game theory and gaming. The complementarity between this approach and the new industrial organization and behavioral theories of the firm is stressed.

Chapter 2 provides an introductory discussion of pure competition and of monopoly. Although these two problems differ considerably, in an extremely important sense they are treated in a similar manner. In particular they are both treated as though the individual, be he a pure competitor or a monopolist, faces a one-person maximization problem. In one instance there is no other competitor who interacts with him. In the other instance, the individual views his environment in such a manner that he does not deem it to be worthwhile to take into explicit account his or any other individual's strategic effect on the environment. The relationship between the "partial equilibrium analysis" of pure competition and the theory of general equilibrium is discussed.

Chapter 3 presents a brief survey of oligopoly theory. It is stressed that there has been a very diverse development of the theories of competition.

One is faced with a long, highly mathematical tradition on the one hand and an equally long, purely institutional tradition on the other.

In spite of the newness of game theory, computer simulation, and gaming, it must be stressed that the work here builds in a direct way upon previous work. Modern methodologies and techniques are used, but this is not a radical departure from the previous analyses and methods of the study of oligopoly. For this reason alone, it would be necessary to provide an adequate survey of the previous work on theories of oligopoly. Beyond that it is important to note how several diverse approaches are now beginning to come together. It was no accident that among the previous approaches both mathematical theorizing and institutional description rated high. Both are needed, and the current trend is for them to blend into a mathematical, institutional economics that provides an abstract underlying structure and at the same time is flexible enough to adjust to the addition of institutional detail called for when we wish to consider the nature of competition in a specific industry, such as automobiles, steel, or oil.

Chapter 4 is devoted to a study of the concept of solution and introduces a certain amount of simple game theoretic notation. It is shown that the specification of the payoffs to each player is tantamount to the specification of the market structure and the goals of the firms. A solution concept may be regarded as the specification of the intents of every player. The solution is the outcome resulting from the application of these intents to the market structure. It is stressed that there is no neat unique meaning that can be given to the concept of rational economic or optimal behavior when there are several individuals with overlapping and diverse interests. Many different concepts of solution to an "*n*-person game" are discussed and related to the study of oligopolistic competition.

Chapter 5, which begins the first formal model building, investigates the structure of a symmetric market model with duopoly without product differentiation. This is the classical case of Cournot and also leads to some of the models of Edgeworth and Bertrand. In Chapter 6 a model of duopoly with product differentiation is intensively investigated. This model is crucial to much of the subsequent work. In particular it is necessary to devote considerable time to expositing the basic problems encountered in the study of oligopolistic demand. The duopoly model presented here is a natural extension of the work of Chamberlin.

The difference between this work and the treatment of Chamberlin comes in the explicit analysis of contingent demand. This enables us to link in a far more direct manner the work of Chamberlin with that of Cournot and Edgeworth. Furthermore, the calculations made are necessary to examine the relationship among stability, capacity, and inventories in a market.

In Chapter 7 a symmetric oligopolistic market with product differentiation is investigated. The assumption of symmetry is obviously a highly restricting one. But although the price paid is high, it yields returns. The mathematical treatment is considerably simpler than it is for the nonsymmetric model. However, certain concepts that are relatively easy to define for a symmetric market are not at all easy to define for a nonsymmetric market. Given symmetry, we are able to suggest an answer to the question of "how many is many" in a market and are able to cast some light on the problem of how competitive a market structure is. These answers do not generalize easily to the nonsymmetric case, but they do provide us with a starting point for such an analysis.

By studying the limiting games obtained by setting n (the number of players) equal to 1 or permitting it to become arbitrarily large, the relationship between monopoly behavior or strictly competitive behavior and oligopolistic behavior is made clear. The former two may be regarded as limiting forms of the latter.

Chapter 8 extends the model to take into account demand fluctuations and the effect of such fluctuations on inventory and capacity. A somewhat surprising result is obtained. It can be shown that under certain circumstances the presence of a fluctuation in the market increases rather than decreases the stability of the market.

Chapter 9 is devoted to the mathematical and economic analysis of a nonsymmetric oligopoly model. This model is a modification of the symmetric game where n-1 new parameters are introduced as "weights" on the n-firms, representing their "intrinsic market share" at a particular period of time. We are able to show that it is possible to obtain solutions for almost all of the solution concepts previously discussed for this particular market model.

It is obvious that in order to develop a satisfactory analysis of oligopoly it is necessary to be able to analyze nonsymmetric markets. The nonsymmetric market model built here is fairly simple, but it is complete and it is amenable to mathematical analysis. With it we are able to at least commence an investigation of the effect of a dominant firm in an oligopolistic market. Furthermore, we can reconsider the measures of the percentage held by the top four firms and the percentage held by the top eight firms.

Although it may be relatively easy to obtain an index for "fewness" when an oligopolistic market is symmetric, it is considerably more difficult to do so when we are confronted with a nonsymmetric structure. We feel that there are some fruitful possibilities to be obtained from the calculation of tables for different types of quadratic market structures.

We also note that with the symmetric game the concept of "going to a limit," that is, of studying the change in the solutions as the number of players becomes large, was well defined. When we begin with a nonsymmetric market this is no longer the case, and the meaning of the growth of the number of firms in the market cannot be dissociated from their sizes and changes in relative size.

Chapters 10 and 11 are devoted to an investigation of advertising and other forms of information for product differentiation. When is advertising advertising? The concept is ill defined, as are the statistics published on the amount spent on advertising. Under-the-counter rebates that are adjustments to price may easily be smuggled into the advertising budget. We give a brief survey of some of the discussion defining advertising and evaluating the pros and cons. We then raise the problem of how the effect of advertising can be modeled to be consistent with other economic theory. For example, should we treat the effect of advertising as a shift in parameters in a well-defined utility function, or should we regard advertising as something that makes it easier for an individual to buy a product but that does not affect his initial preferences? The remainder of our work on advertising is devoted to investigating a specific functional form applied to an oligopoly model and to deriving the various solutions.

This is a book on economic theory, yet it is constructed so that virtually all of the theory developed can be illustrated and further investigated by means of a market or business game that can be played on a computer by one or more individuals. The game is designed as an aid to teaching, a vehicle for research and experimentation, and a preliminary model for operational use. The book stands by itself and in no way depends upon the decision to use the game. Nevertheless the game is of considerable value in illustrating the economic theory and in giving the reader a more vivid impression of the link from theory to institutions, the flow of information, and the dynamics of competition than is provided by mathematical analysis alone.

Documentation manuals and several computer programs for this game exist. They can be obtained from Social Systems Incorporated, 300 Eastowne Drive, Chapel Hill, North Carolina 27514.

The use of a game as a link between economic status and corporate dynamics is illustrated in translating the information generated by the economic model into the format used in corporate decisionmaking. In Appendix A, on accounting and economic theory, the uses of the game more or less coincide with the exposition of theory and the subject matter of accounting. The various output sheets obtained from playing the game are displayed. They contain the balance sheet, profit and loss statement, and competitive market information. Every line on each of these outputs is discussed and the concept examined. In other words, there is a complete linking of these basic accounting items to the oligopoly model developed previously. This appendix provides information on the details of the game, but also bridges the gap between information as it appears to the individual in business, in this case the player, and the economic model and market abstraction as it appears to the economic theorist. The link between economic theory and accounting is of considerable importance to the economic theorist in general and the student of industrial organization in particular. Much work is usually needed to translate legal and accounting profit measures into economic measures.

Appendix B contains the general theory of contingent demand and indicates its relationship to the problem of rationing. This is somewhat more technical than the text and can be omitted by those who have understood the basic functioning of the market mechanism under oligopolistic competition as illustrated in Chapter 6 for the case of duopoly with differentiated products. Those interested in the more technical aspects of rationing and of algorithms designed to allocate demand under conditions of various price and capacity constraints will find this appendix of interest.

We wish to thank the following journals and presses for granting us permission to use some of our articles: *Kyklos* for "Information, Duopoly, and Competitive Markets: A Sensitivity Analysis" and "Price Strategy Oligopoly with Product Variation"; *Western Economic Journal* for "Price Strategy Oligopoly: Limiting Behavior with Product Differentiation" and "A Further Comparison of Some Models of Duopoly"; *International Economic Review* for "Price Duopoly and Capacity Constraints"; *Journal of Economic Theory* for "Price Variation Duopoly with Differentiated Products and Random Demand"; *International Journal of Game Theory* for "Duopoly with Price and Quantity as Strategic Variables"; and North Holland Press for "Noncooperative Equilibria and Strategy Spaces in an Oligopolistic Market," which appeared in H. W. Kuhn and G. P. Szego, eds., *Differential Games and Related Topics*.

Part of the research preliminary to this book was supported by the Office of Naval Research, Contract Nonr-3055(00) and subsequent contracts.

We also wish to acknowledge the valuable discussions and collaboration with several colleagues, including Lloyd Shapley, Gerrit Wolf, Austin Hoggatt, Matthew Sobel, Vernon Smith, and Reinhard Selten. We wish to thank Elizabeth Walker and Glenna Ames for typing most of the many versions of this manuscript.

Contents

Figures

Tables

Market Structure and Behavior

1 | Introduction

COMPETITION AMONG A FEW FIRMS is not restricted to a single topic in the study of economics, but poses many problems across disciplines: both structure and behavior must be studied. The traditional economic variables of price and quantity of product play an important role in the study of competition or collusion, but they are only two among a host of variables.

Different industries have different problems. This does not exclude generalizations, or the construction of worthwhile theories; however, the path from general theorizing in the economics of oligopoly to the study of specific industries involves far more attention to detail and the specification of special variables relevant to each case than is customary in traditional economic study. Among the features of considerable importance in determining the nature of competition are price, amount produced, product differentiation, variation of product differentiation, capacity constraints, advertising, distribution, production scheduling, inventory costs, costs of changing production, plant expansion, entry and exit conditions, ruin and bankruptcy conditions, ability to borrow funds, liquidity, corporate form, internal organization, stockholding conditions and dividends, regulation, and antitrust. Special institutional circumstances may play a vital role beyond the technical and specifically economic and financial conditions. Special patents, secret formulas, a well-known and trusted name, habit, sentiment, and privilege can be keys to understanding a specific industry.

When firms are few and individual units large, the convenient fiction of treating each as the basic unit of microeconomic theory obtains only limited results. The behavioral and organizational aspects of the structure of the firm are obviously of interest. Multiple product and plant firms are faced with problems of centralization and decentralization that require an understanding of the management, ownership, and employees, their sociopsychological relationships, the legal structure in which they operate, and the total societal environment of the firm. Natural and implicitly accepted norms in one society may be utterly foreign to another and these differences may be manifested in the organization of the firms.

A large bureaucracy operates in a labyrinth. Information is often difficult to obtain and frequently informal lines of communication are more important than official channels. The goals of the individual employee are not necessarily those of the organization as a whole; indeed it may be impossible to describe the goals of the whole organization. Many individuals within the organization wish to reduce uncertainty and live the quiet life; others seek a more heroic existence fancying themselves ever-fresh entrepreneurs and concentrating on the search for new products, processes, or markets.

Behavioral and bureaucratic theories of the firm are needed to explain the multifaceted nature of the large corporation. A start has been made in such works as those of Bonini (1963), Cyert and March (1963), Arrow (1974), Nelson and Winter (1974), Williamson (1975), and others. The new industrial organization must put flesh on the bones of theory (wherever those bones can be found to exist). The works of Williamson and of Nelson and Winter among others go a considerable distance in putting the detailed structure of industrial organization into context. But we must keep in mind that detail in modeling is highly linked to the questions that need to be answered. Oligopoly theory in various manifestations may be of use to policy oriented bureaucrats concerned with control of industry, lawyers, oligopolists, mathematical economists, and others. But different stresses must be given for different purposes.

The goals for this book are modest. The hope is to bring classical oligopoly theory and mathematical economics closer to the new work in industrial organization and to promote an interest in simple game experiments as a way of increasing our understanding of the meaning of competitive and cooperative behavior and structure.

How are competitive decisions made in the large corporations? How important are competitive decisions as compared with the actions involved in running the internal processes of the firm? Much of the theorizing about oligopolistic behavior has been devoted to sketching purportedly reasonable but noninstitutional processes of competitive or collusive interaction. From Cournot (1838) onward the literature is filled with suggestions for the actions and reactions of firms whose forces influence the fate of one another in the marketplace. In recent years game theory and decision theory (von Neumann and Morgenstern, 1953; Luce and Raiffa, 1957; Shubik, 1959b) have been added to the tools of analysis. Due to the influence of operations research, the computer revolution, and the modern business school, many executives are beginning to talk and conceptualize in terms of explicitly perceived models of competitive structures (Schrieber, 1970).

Markets can be studied in many ways and many different questions can be posed, most of which are relevant to policy problems. A single book cannot possibly cover the multitude of alternatives; in narrowing the scope of inquiry several drastic simplifications must be made. Specifically, much of the analysis developed in this book (and most of the formal theorizing on oligopoly problems) uses the techniques of partial equilibrium analysis. The full feedbacks and reactions from society on the firm are played down. External economies or diseconomies and various important noneconomic values are only observed incidentally.

Structure or Behavior

Is society concerned with results and actions in the marketplace or with intent? Is bigness, or should it be, a crime? How many competitors is many? What is the economic price paid for breaking up large firms or preventing mergers? These are all critical questions and the answers involve a mixture of value judgments and analyses. Setting aside our value judgments we propose to concentrate on the investigation of the relationship among intent, behavior, and market structure.

Let us assume that the intents of a group of individuals are the same. Can we give operational meaning to measuring the influence of structure upon their behavior? In other words, can we illustrate the nature of the behavioral change of individuals with the same intent as surrounding circumstances change?

The methodology of game theory provides a new and useful approach to studying the interrelationship between structure and behavior. It casts new light on concepts such as conscious parallelism, implicit collusion, signaling, and the design of policing systems. Although game theory may be viewed normatively or behaviorally, the tendency has been to consider it more as normative than behavioral theory. One reason for this has been that in the behavioral sciences, particularly economics, the observer has very little control over the phenomena being observed. General theory is often well concealed by a host of special institutional facts especially when studying behavior of individual institutions.

Occasionally clear evidence of specific behavior in a sufficiently well defined and simple structure can be obtained. For instance, evidence comes to light on prearrangements concerning the submission of sealed bids and both the economist and the lawyer are in a position to test their theories with a fair amount of confidence. In general, however, information on behavior and market structure is complicated and costly to obtain.

A Unified Variety of Approaches

With a topic as complex as market structure and behavior there is a temptation to spread the analysis thin as many different fascinating but only partially relevant subjects are pursued. In contrast, the purist may wish to lavish analysis in depth on only a single aspect of the problem of competition. In this volume we attempt to reach a compromise.

The central theme is the study of competitive behavior and the structure of markets. The major methods used are those of economic theory, the theory of games, simulation, and gaming. These, however, are not sufficient; other topics must be blended in. Subjects such as finance, marketing, and accounting that have often been regarded as stepchildren of "serious economic theory" cannot be ignored. Furthermore, topics recently developed under the more technological auspices of operations research, such as inventory theory and production scheduling, have considerable influence upon the structure of industry and cannot be dismissed as merely technical details that can be adequately handled by engineers.

Things in general, and economic things in particular, are seldom what they seem. The microeconomic theorist or the lawyer who fails to appreciate the Alice-in-Wonderland art of corporate reporting fails to be proficient in his profession:

> "When I make a word do a lot of work like that," said Humpty Dumpty, "I always pay it extra."

The operational problems of the firm are not only technical, economic, financial, and legal; they are behavioral as well. Individuals must coordinate; they must get along with each other: trust is needed. The environment undoubtedly contains uncertainty. Information is a precious commodity and distortions in information abound. The scope of this volume does not permit us to do justice to these topics. We do not dismiss them, but attempt to note where and when their relevance is greatest. This occurs whenever we turn our attention to the development of a dynamic theory.

2 | Monopoly and Pure Competition

AS IS CUSTOMARY in much of economic analysis, we use diagrams as well as mathematical analysis. The art and science of using diagrams for illustrative purposes is possibly more highly developed in economic theorizing than in any other subject of study. For many decades economists have recognized that when one wishes to indicate a direction or a tendency without necessarily needing numerical accuracy, diagrammatic methods are more flexible and understandable than mathematical methods. Unfortunately this is not the case when three or more variables are involved. Furthermore in certain situations, "how much" becomes a vital question calling for mathematical models and methods. However, at the outset we utilize the traditional expository tools and present a diagrammatic and mathematical representation of monopoly and pure competition.

Monopoly and Monopsony

A firm is monopolistic if it controls the sale of any particular good or service to the complete exclusion of other firms. It is monopsonistic if it is the only institution in a position to buy a specific good or service. Obviously it is possible to be monopolistic or monopsonistic with more than one item. In many countries the central government plays this role. As with most categorizations in the behavioral sciences the category of monopoly is one of degree. The usefulness of the concept depends on our understanding of the relevance of the degree of the definition. The more close substitutes that exist for a given product, the less meaningful is a discussion of the role of a firm as a monopolist since it is selling a uniquely discernible product that is easily replaced. It is unlikely that the monopoly power conferred upon an individual as the sole supplier of vanilla ice cream containing 263 chocolate dots for every pound sold will be of much concern to either the economist or the antitrust lawyer. On the other hand, an individual who managed to obtain a monopoly on salt or pepper would be in a much stronger position in the market!

Some monopolies are inherent or natural: there is only one Yosemite Valley and the talents of Charlie Chaplin were basically unique. The forces of technology may create a monopoly in certain industries: often one railroad or power generation station is all that is needed for a district. Monopolies are also created by law, custom, and individual force.

The expression, monopolistic competition, is used (Chamberlin, 1962) to account for the transition zone at which firms are selling differentiated products; in fact, it may be a source of great pride and profit to stress brand names. However, one bar of soap is not so different from another that one firm has unlimited power in the market. The central issue in monopolistic competition is that of interaction among the firms. (Discussion of this phenomenon will be delayed until the problems of isolated, individual behavior have been addressed.)

Absolutely pure monopoly does not exist in the sense that any good or service is competing with all others as a substitute for the use of the consumers' income. For the purpose of our analysis, we need not become enmeshed in the philosophical niceties of this point. To a very good first-order approximation, monopolies can and have existed. When the level of substitution between a specific uniquely held product and any other commodity is low, the individual holding the product is in a position to exercise considerable market power. Very often exercise of this power takes the form of decisions regarding price. Specifically, with some idea of the nature of the product demand, an individual is able to maximize profits by a judicious selection of price combined with control over production or supply. Monopolies on salt, tobacco, liquor, gambling, and drugs, for example, all manifest considerable market power. Economic history illustrates that this lesson was fully understood when various governments decided to select the state monopolies used to raise revenues.

Our use of the word monopoly in this chapter is limited to the situation in which a single firm is selling a product for which substitutes are sufficiently distant that the actions or reactions of potential competitors need not be considered. Furthermore, the market contains sufficiently large numbers of small buyers that the firm may regard its demand as being given by an impersonal overall behavioral equation called the market demand. In other words, the strategic actions of the individual buyers are basically irrelevant; they have no power and therefore no strategic influence in the marketplace.

In the theory of games, these observations amount to specifying that there is no other player in the market on the seller's side and that the customers are not regarded as strategic entities but have merely been approximated by a mechanism. The monopolist is faced with a one-person maximization problem. These are the givens even though legal, social, and political mechanisms of society may restrain the actions of the monopolist in the long run. In fact, if a strategic struggle of any variety is encountered, it is more likely that the "players with power" are not in the market at all, but are legal or political entities.

The economic problem faced by the monopolist is one of simple maximization. If he sells many monopolistic products and faces an uncertain market or has complicated production processes, then, even though conceptually a straightforward maximization problem is faced, it can be extremely difficult to solve. In this chapter, however, we limit ourselves to the single-product form in a static context, in which uncertainty, production scheduling, inventory problems, costs of changing production, and other features making the selection of an optimal policy a difficult intellectual problem remain unexamined.[1]

With a single product, the maximization problem can be easily illustrated by means of a diagram. This is shown in Figure 2.1. There are four curves of importance to be noted: The first, labeled DD', is the demand curve for the product showing the relationship between the price charged and the amount that can be sold at that price; or alternatively, the price that would be needed to clear the market if the demand is to equal a specific supply. This relationship is expressed

$$q = f(p) \quad \text{or} \quad p = \phi(q), \tag{2.1}$$

depending upon which variable we consider to be independent. This equation holds true only if the demand curve does not bend back so that a vertical line could cut it more than once.

The second curve of importance, labeled MR, is the marginal revenue curve showing the relationship between changes in revenue and changes in sales. Assuming that $\phi(q)$ is continuous and differentiable, total revenue is expressed as

$$qp \quad \text{or} \quad q\phi(q).$$

1. The reader interested in input mix problems should refer to a text on microeconomics or programming such as Henderson and Quant (1971) or Charnes and Cooper (1961).

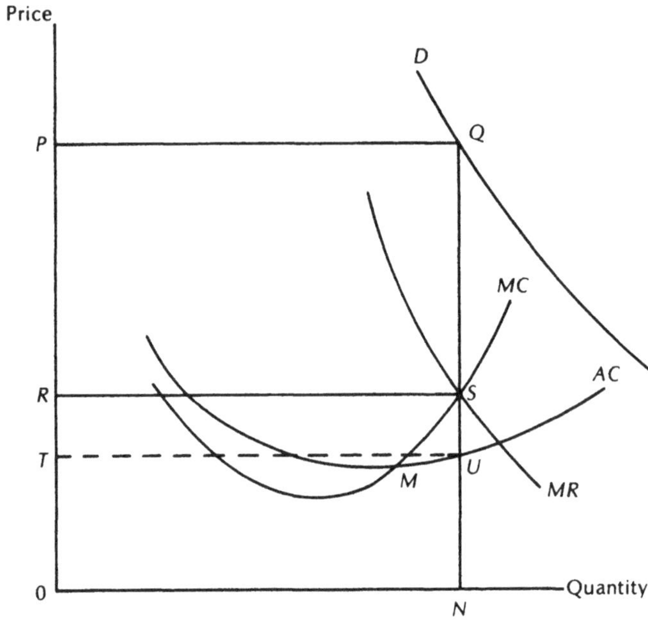

Figure 2.1. Monopolistic maximization.

By differentiating the condition for total revenue, we obtain the marginal revenue curve:

$$\phi(q) + qd\phi(q).$$

The curve labeled *AC* represents the average cost curve of the firm. The curve starts to fall, reaches a minimum and then increases. The argument for this U-shaped curve is that when the firm is operating initially at production rates below its efficient size it has unused plant and overheads leading to extremely high average production costs. As production increases so does the efficiency of the plant until at some point an optimum size is reached for this particular combination of equipment, management, and variable materials. Beyond that point bottlenecks and inefficiencies force average costs to rise again. The average cost curve is generally drawn with no beginning and end points. However, extending the average cost toward zero production creates difficulties: at zero production the average by definition will be infinite unless a more sensible mathematical model is derived. In general it is reasonable to assume that there will be some minimum production level below which the firm cannot go; for example, it is unlikely that even the most enterprising of small sports car manufacturers would contemplate a production level of three-quarters of an automobile.

There is also no indication of how the right-hand side of the curve would behave if production were considerably increased. Since this is a static representation of costs, it is assumed that a capacity limitation would be reached that could only be overcome after sufficient time had passed. The average cost is expressed mathematically as $C(q)/q$, where $C(q)$ is the total cost of production of q units.

The marginal cost curve can be drawn in a manner analogous to that of the marginal revenue curve showing the change in the incremental cost of producing an extra unit for each level of production. If we assume that the cost curve is continuous and differentiable, the marginal cost curve, obtained by taking the first derivative of $C(q)$, is expressed as $dC(q)/dq$. It can be easily shown that the marginal cost curve, labeled *MC* in Figure 2.1, intersects the average cost curve at its lowest point. Taking the first derivative of $C(q)/q$ and setting it equal to 0, we obtain

$$\frac{-C(q)}{q^2} + \left(\frac{dC(q)}{dq}\right)\frac{1}{q} = 0, \tag{2.2}$$

which becomes

$$\frac{dC(q)}{dq} = \frac{C(q)}{q}, \qquad \text{for} \quad q > 0. \tag{2.3}$$

A straightforward consideration of the maximization process tells us that the monopolist will obtain the greatest profit when price is selected such that marginal cost equals marginal revenue. When marginal revenues exceed marginal costs any incremental unit adds to profit. This addition ceases when the curves intersect. In Figure 2.1, the point of intersection is indicated by *S*; the price charged by the firm is *OP* and the amount sold *PQ*; total revenues are indicated by the rectangle *ONPQ*, total costs by the rectangle *OTUN*. Therefore, the profit is indicated by the rectangle *TPQU*. We are assuming that the demand curve is of a regular shape; that is, it slopes downward to the right implying that as price decreases, sales of the product increase. While certain inferior goods provide an exception to this rule, they are not considered at this point.

This result could have been obtained using the revenue equation for the firm:

$$\Pi = q\phi(q) - C(q). \tag{2.4}$$

Taking the first derivative of this expression and setting it equal to 0, we have

$$\phi(q) + q\,d\phi(q)/dq = dC(q)/dq \tag{2.5}$$

or marginal cost equals marginal revenue.

Suppose that the demand faced by a monopolistic firm is given by

$$d = \alpha - \beta p, \tag{2.6}$$

where d equals demand and p equals price.

Furthermore suppose that the firm has constant average costs given by

$$\frac{C(q)}{q} = \frac{cq}{q} = c, \tag{2.7}$$

where q is the amount produced. Figures 2.2 and 2.3 show the geometry of sales, revenues, and costs and its relation to the analytical method somewhat differently than does Figure 2.1.

The total revenue curve TR in Figure 2.2 can be expressed algebraically as either

$$TR = p(\alpha - \beta p) \quad \text{or} \quad q\left(\frac{\alpha - q}{\beta}\right).$$

The second form for TR is obtained by inverting the demand function, or expressing p as a function of d:

$$p = \frac{\alpha - d}{\beta}, \tag{2.8}$$

q is then substituted for d. This amounts to saying that demand will be the same as supply, which is the same as production in a static or "timeless" market.

$$\frac{dTR}{dq} = \frac{dq(\alpha - q)/\beta}{dq} = \frac{\alpha - 2q}{\beta}.$$

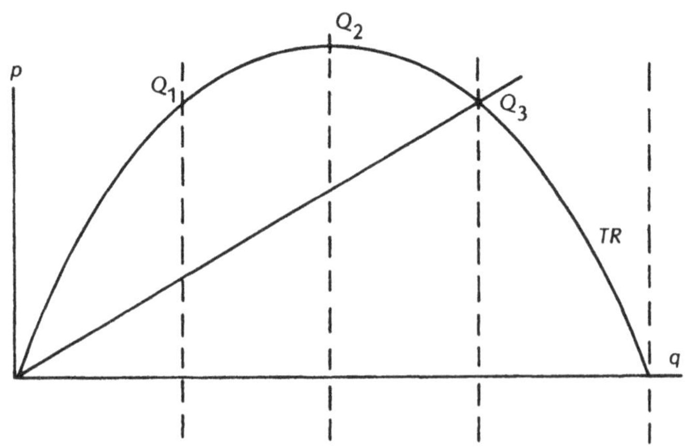

Figure 2.2. Total revenues.

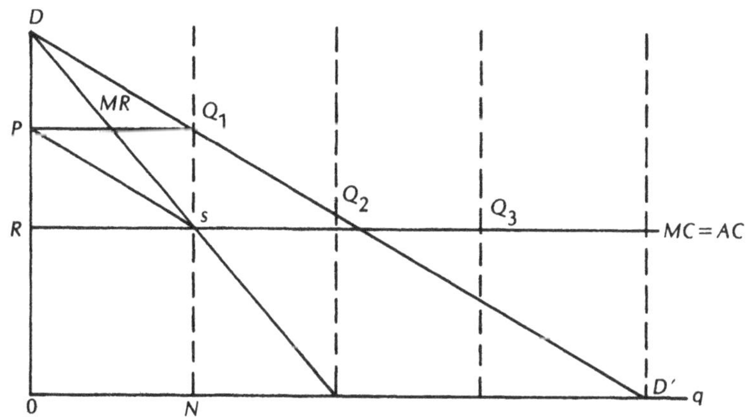

Figure 2.3. Monopolistic maximization: linear demand.

This is the marginal revenue curve *MR*, which is drawn in Figure 2.3. It has twice the slope of the demand curve.

Here we make the simplifying assumption that only one aggregate input enters the production process. That is, we implicitly assume that the optimum mix of inputs has been selected. The total cost curve *TC* is given by

$$TC = cq,$$

hence, the marginal cost is

$$MC = \frac{dTC}{dq} = \frac{dcq}{dq} = c \quad \text{and} \quad p = \frac{\alpha + \beta c}{2\beta}. \tag{2.9}$$

This equals the average cost in this case. The monopoly point is given by Q_1 in Figures 2.2 and 2.3.

Pure and Perfect Competition

The contrasting state to pure monopoly, which is equally unrealistic, is pure competition. If we wish to heighten the lack of realism we might refer to it as perfect competition. Speaking loosely, the general idea behind the concept of a purely competitive market is that there are many competitors, each selling an identical product. It is plausible that in this circumstance if the products being sold are genuinely identical then the tyranny of the consumer will reign. A consumer will prefer to pay less for the identical item if the effort involved in obtaining it is the same in all instances. This means that in contrast with the monopolist who had control over price, the competitor in a purely competitive market will have virtually no individual control over the price since the firm who names the lowest price inherits the market.

Price will be driven down to the cost of production of the marginal active high-cost producer. This does not mean that the competitor will starve, because added into the costs of production are wages or the value of whatever services are performed in the production process. It is paradoxical, but good economics to speak of the pure competitor obtaining no profits in the competitive market, in the sense that the competitor is rewarded for the worth of the particular production services and no more. In other words, the competitor is able to command no higher a reward than could be commanded if he closed his own operation and earned a wage from someone else. (In Appendix A we return to the study of the meaning of the word profit as it is used in economics, business, taxation, and accounting.)

The difference between pure competition and perfect competition is made here only to guard against confusing economic theorizing with bad philosophical speculation. To conceptualize the decision problem faced by the individual entrepreneur or firm in a competitive market we must describe the relationship among the products and the knowledge of the individual competitors. Are the products genuinely identical to the extent that the bother of buying from Ⓐ is absolutely the same as the bother of buying from Ⓑ? In reality, this is never the case. Do the individual competitors have the same base of information? What do they know about their costs, their competitors' costs, and the market demand? In a market having perfect competition we make the unrealistic assumption that the individuals have absolutely identical products. Furthermore, they know their own costs and their competitors' costs in complete detail; entry and exit from the market is absolutely free and frictionless; the demand for the product they are selling is completely known.

Some writers have gone further to suggest that a competitor in a perfect market also has perfect foresight; not only does the perfect competitor know absolutely everything about today's price of beans, he is absolutely informed about the price of beans tomorrow and all days after. There are several unsatisfactory philosophical issues that would have to be resolved if we seriously thought that individuals required perfect foresight to operate in an economy. Fortunately, we do not think that this is so and hence even for the limiting instance, we address ourselves to pure competition meaning that as a *reasonable first approximation* to this kind of market we can assume that, for the most part individuals are selling approximately the same goods. In other words, taking into account the additional costs of transportation and other convenience factors a customer has no intention of paying more for any one of a group of goods than the lowest price, corrected for a few convenience factors.

As for the information base, individual sellers are assumed to know their own costs, entry is easy, no individual has any control over price since all products are close substitutes, and everyone knows the current market price. If we wish to introduce dynamics and worry about future prices, we need not make every competitor a universal genius but we assume that he has some expectation of what the price of wheat, bananas, or baseball bats will be a year hence.

A corollary to this model is that selling costs, fancy retailing, advertising, and other forms of product differentiation do not matter in the competitive market. Chicken soup is chicken soup whether it is Mother's chicken soup or Marilyn Monroe's chicken soup; the product is homogeneous and brand names are of no importance. For most items this is not the way things are, but for our purposes, let us imagine that we do not care whose brand of salt, lead pencils, or paper napkins we buy.

Actually, when one stops to think about it almost all consumer products are differentiated. The only area in which product differentiation does not matter is probably industrial supplies where when one wants lead, tin, or zinc of a certain quality, the product of one producer is a pretty good approximation of the product of the other. Paradoxically, precisely in such cases, the supply is usually oligopolistic and although the physical products are homogeneous, the supply conditions, such as timing and special aspects of delivery, are often not the same.

We have treated monopoly as a useful extreme of competition. We might as well balance the discussion by treating the other end of the spectrum, the extreme of pure competition.

From the viewpoint of studying decisionmaking processes, the state of pure competition presents a simple and striking paradox. With pure competition individual control in the marketplace is so weak that one might as well forget that one is a competitor. Rather than viewing the market as though it contained myriad competitors against whom the individual matches wits and skills, it becomes a vast impersonal system and the competitor is seen as a decentralized cog in the system with invisible people guided by an invisible hand. The real paradox is why economists have chosen to call this state of affairs competitive. In common usage competition suggests competitors who are individually aware of each other's presence. In this particular case, the apotheosis of competition is reached in which no individual competitor matters to any other individual competitor!

The idea of individual behavior in a market with pure competition can be easily illustrated in a diagram as shown in Figure 2.4. The demand curve to the individual competitor appears to be a horizontal line. In fact, it is really not quite horizontal but the individual competitor is assumed to be small relative to the overall size of the market, and as a good first approximation, one can assume that what the competitor offers the market will have no influence on the market price. Here we have a plain

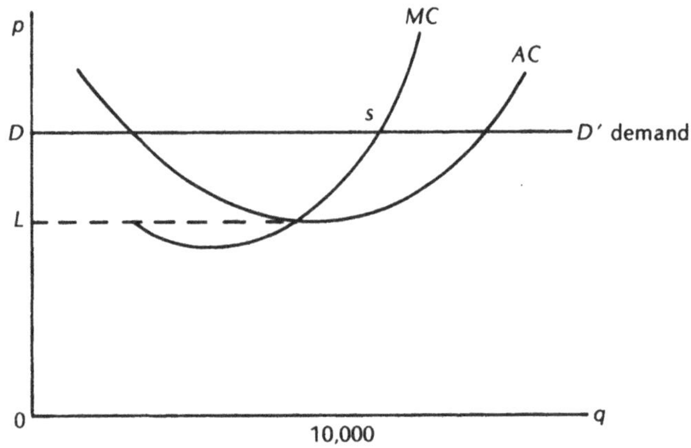

Figure 2.4. Profit maximization in pure competition.

case of fallacy of composition: that no individual influences market price is undoubtedly true; however, it does not follow that the actions of all individuals taken as a whole have no influence on the market price. If we were to plot the total demand curve for the product and the total supply evoked from all firms at different prices, we would get the diagram shown in Figure 2.6, which by no means has a horizontal demand curve.

Since $TR = pq - C(q)$,

$$p = \frac{dC(q)}{d(q)} \quad \text{and} \quad \frac{d^2C(q)}{dq^2} \geq 0 \tag{2.10}$$

at a maximum. At equilibrium in a competitive market a firm sets its production at the point where marginal costs equal price. This is shown as the point s in Figure 2.4.

If we assume that average costs of production are constant, say $C(q) = cq$ then $C(q)/q = c$. If c is greater than market price the firm will produce nothing; if c is less than or equal to market price the optimum production for the firm is limited only by the firm's ability to produce. It is reasonable therefore to introduce a capacity limit k as is shown in Figure 2.5.

From Figure 2.4 and Eq. (2.10) we see that for any price above L the firm will supply the amount indicated by the curve MC (below L price is lower than lowest average cost and it does not pay to produce). To obtain the industry supply curve we add the potential supplies indicated by each marginal cost curve. This results in the supply curve $\overline{ss'}$ in Figure 2.6. Similarly if we add the (apparently horizontal) demand curves we obtain the aggregate demand curve $\overline{DD'}$ shown in Figure 2.6. The point s is the market equilibrium price. In Figure 2.4 we suggested production capacity of around 10,000 for the individual firm at equilibrium. In Figure 2.6 in-

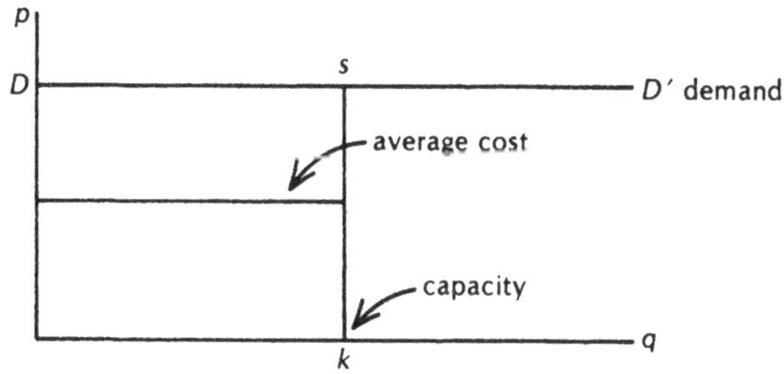

Figure 2.5. Fixed average costs with limited capacity.

dustry sales at equilibrium are given at 1,000,000, thus the slope of DD' although apparently zero is in fact 1/100th of the slope of $\overline{DD'}$, which as a first approximation can be treated as horizontal.

Comparing Monopoly and Pure Competition

Monopoly and pure competition, while utterly different, share two important characteristics. They both present a one-person maximization problem and in each the concept of an industry is well defined.

The monopolist looks at his demand curve, a line generally sloping downward toward the right, and calculates optimum price and production policy. The pure competitor looks at his impersonal demand curve, which to a good first approximation is nothing more than a horizontal straight line, and realizes that he has no influence over price but must nevertheless pick an optimum production schedule. Both individuals need not take into account the actions or reactions of others.

We may conceive of an industry as a group of firms selling the identical product. We exclude other firms if the product they are offering is in any way differentiated. In the case of the monopolist the industry consists of the individual, the only seller of a particular product; hence the definition

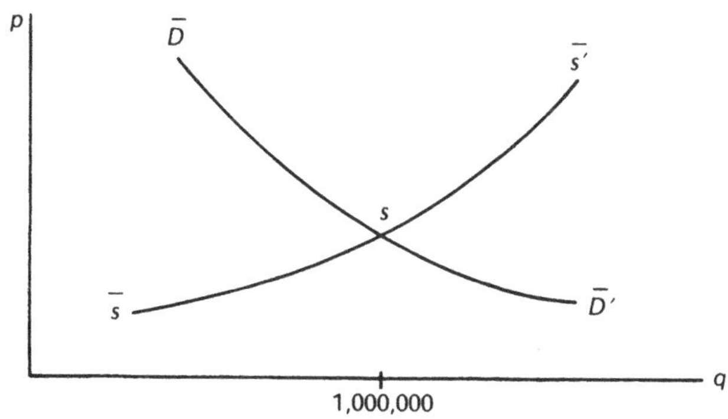

Figure 2.6. Competitive aggregate supply and demand.

holds. In the case of the pure competitor we assume that the other pure competitors are selling the identical product and even if allowances are made for trivial differences, such as location, a bushel of wheat of a certain specified quality is a bushel of wheat and a package of salt regardless of who has ground it is a package of salt. Hence, if we wish to consider the ideal case in which a series of single-product firms are selling the identical product, we can without too much difficulty lump Farmer Jones and Farmer Smith and Farmer Schultz together and say they form the wheat industry.

The concept of industry is difficult to define for multiproduct firms, each of which is selling differentiated products, of which some are selling services, others are selling integrated systems, and all are varying their products from month to month. Even if firms are not varying products, many at least attempt to vary services and spend a great amount of time telling customers how they differ from each other. In a circumstance such as this it is difficult to talk about an industry. This is both an intellectual and a practical operational problem.

The concept of an industry is useful for many different purposes. However, one use should not be confused with another. In some cases, the concept of industry may be completely misleading. For example, in the study of competition, a technological product similarity definition of industry may be misleading in the extreme. Sports cars may be in greater competition with boats and luxury yachts, or small planes, than with family sedans.

The maximization rule derived for the monopolist can be modified to supply the maximization rule for the pure competitor (see Figures 2.1 and 2.4). Whereas marginal cost equals marginal revenue for the monopolist, this rule is modified for the pure competitor to become price equals marginal cost. Both are the same rule, since the demand curve is a horizontal line giving price and marginal revenue the same value. Looking at this rule operationally tells us that the firm has no control over its destiny in the market. The only variable over which the firm has control is internal efficiency. The market in this case is looked at as an efficient decentralizing device. The individual firm tries to maximize its profits with the understanding that price is given.

Figure 2.6 shows the equilibrium state for the industry. The marginal firm, that is, the last firm to consider producing worthwhile, finds that when it enters the market, not only does it set marginal costs equal to price but at that particular point its average costs equal its marginal costs. In other words, the marginal firm makes no profits whatsoever. As previously indicated the economist does not talk about profit in the same sense as its common usage would imply. The fact that the marginal firm makes no profit means that once the owner has paid to himself the wages that he could have earned elsewhere there is no further surplus. "Creative accounting" and the imperfections in the markets for managerial talent enable the managers or owners of marginal firms, in some instances, to subsist in comfort.

Maximizing Sales and Breakeven Solutions

In the folklore of business, corporations may attempt to maximize sales or gross revenues rather than profits. In Figure 2.2 this accounts for selling at the point Q_2. Simple stripped-down economic models of the variety presented in this chapter make it appear foolish for the firm to maximize gross revenues rather than net revenues, or sales volume rather than profits.

However in large corporate bureaucracies where the interests of management are not necessarily synonymous with the interests of the owners, the power of the manager may relate better to the size of the organization and its volume of sales than to profits. Furthermore gross revenues are generally far easier to measure than net revenues whenever cost accounting becomes at all complicated. Although accounting for all costs in static models of single-product firms appears as a simple and straightforward procedure, this is far from true when dynamics, uncertainty, and multiplant corporations are considered.

In the simple model presented in this chapter the maximization of sales amounts to

$$\text{Max } q\phi(q). \tag{2.11}$$

This should be compared with Eq. (2.4) which describes the condition for the maximization of profits.

Another important rule-of-thumb behavior attributed to firms is that they operate on the basis of cost plus mark-up. This is closely related to breakeven analysis in which the first thing the owner of the firm worries about is that given a price, which the owner thinks he can get for his product, will there be enough demand to cover costs or do better. Figure 2.7 provides an example in which K is the fixed cost and total cost is expressed as $C = cq + K$, while $R = pq$ is gross revenue obtained from selling q at price p. The break-even volume needed is represented by q^*. Breakeven analysis is the type of check that a firm unaware of the shape of its demand curve but desiring to estimate the demand needed to operate out of the red would be expected to make.

Fixed costs played a role in the breakeven analysis. They do not play a role in profit maximizing analysis. If we included a fixed cost K in Eq. (2.4) the resultant prices and sales for optimum revenues would not be changed. Why then should the business rule-of-thumb differ so from the economic maximization rule? The answer is that they exist on different

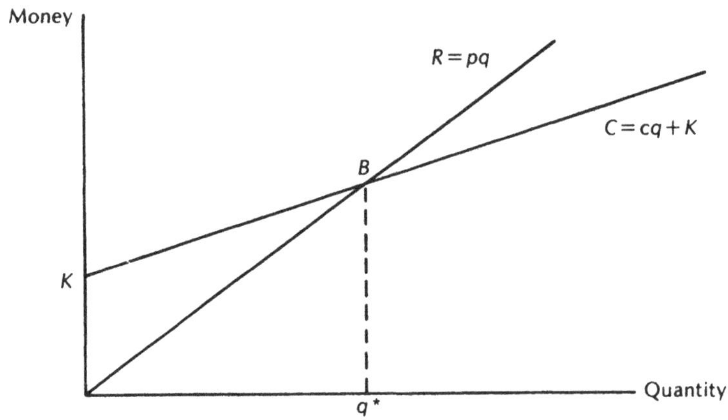

Figure 2.7. A breakeven chart.

time scales. The economic analysis is short run in the sense that the fixed costs are regarded as sunk. The breakeven analysis is the risk calculation to be made before committing oneself to fixed costs.

It is important to remember that at best economics is an applied science and all of its models are some form of approximation. In particular a fixed cost given one set of actions and one time horizon may become a strategic variable when considering a different set of activities and another horizon.

Further Comments and a Simple Model

The monopoly and pure competition models involve one-person maximization. Here the problems appear almost trivial. In actuality the firm faces uncertain demand, detailed production problems, and a host of details as yet undiscussed. Even at this level of simplicity the reader may wish to experiment with a monopolistic or competitive market model. We use the following parameter values in this and subsequent chapters for purposes of comparisons:

$$\alpha = 1{,}000{,}000$$
$$\beta = 1{,}125$$
$$c = 200.$$

The following table shows three solutions:

Solution	p	q	P
Monopoly	544.4	387,500	133,472,222
Maximum sales	400.0	500,000	100,000,000
Pure competition	200.0	775,000	0

The first solution associates price, production, and profits with the monopoly price. In the second solution sales or gross revenues rather than profits are maximized. The third solution shows the market size under pure competition when $p = c$. The solution points are indicated in Figures 2.3 and 2.4 by the points Q_1, Q_2, and Q_3. The following table shows the mathematical expressions associated with the solutions:

SOLUTION	p	q	P
Monopoly	$(\alpha + \beta c)/2\beta$	$(\alpha - \beta c)/2$	$(\alpha - \beta c)^2/4\beta$
Maximum sales	$\alpha/2\beta$	$\alpha/2$	$[\alpha(\alpha - 2\beta c)]/\beta$
Pure competition	c	$\alpha - \beta c$	0

Accounting Taxes and Overheads

Most firms pay taxes, have overhead costs, and report economic activity in documents that conform to accounting procedures. Generally Accepted Accounting Principles (GAAP) provide operational rules for making approximate measures in money of economic magnitudes in a dynamic context. The treatment of overheads, advertising costs, costs of changing production, inventory carrying costs, depreciation and so forth requires the use of accounting conventions to provide consistent methods for gathering and reporting economic information. The mathematical and diagrammatic models should translate into accounting form to fully appreciate the various microeconomic models of the firm. This is done in Appendix A for the models presented here.

The economic rules for profit maximization appear straightforward when applied to the simple stripped-down models analyzed here. However, when taxation, corporate form, and aspects of dynamics and uncertainty are taken into account, the rules become complex in application even though the principles remain unchanged.

3 | Theories of Oligopoly

THE CHALLENGES posed by the study of oligopoly are many for its problems are complex and formulating with precision many of its basic questions can be difficult. How is the level of collusion to be measured in a market? What are meaningful indices for the measure of market power? When does intensive competition become "cut-throat" competition? What criteria should be used to determine whether an industry is a natural monopoly that should be regulated? What are the social losses or gains from oligopoly? How do attempts to preserve competition relate to the preservation of competitors? Obviously if all competitors but one fail to survive there no longer is any competition; however, the intent of the laws for preservation of competition may not be fulfilled by preserving inefficient competitors. Here we have an example of a quantitative change bringing about a qualitative difference. At least two surviving competitors are needed to maintain competition in a market yet such efforts to preserve nonviable competitors may actually destroy rather than preserve the level of competition.

The control of industry and oligopolistic competition, an important societal concern, is difficult to study empirically and poses logical and analytical problems. Perhaps it is for these reasons that great dissatisfaction with the state of oligopoly theory exists at this time. The nature of the problems besetting any attempt to construct a theory of oligopoly or to render useful guidance for policy and control has led to the development of two different approaches to the subject. They are the mathematical and institutional approaches. There is a history of mathematical models of oligopolistic competition dating from Cournot (1838) to the theory of games. There is also a literature generated by institutional economists, lawyers, and administrators interested in formulating and implementing public policy. It has been the tendency of these groups to work almost as though the other did not exist. Despite lip service rendered to most of the theoretical writings in oligopoly theory it is doubtful if the theorizing has been of more than incidental value to the formulation of policy. It is the thesis of this book that both the mathematical and the institutional approaches to oligopoly were natural and necessary as preliminaries to an approach needed to combine the degree of abstraction stressed in the mathematical approach with the concrete recognition of institutional facts and details emphasized by those directly concerned with the control of industry.

Prior to the 1930s, the institutional and mathematical approaches were the two main streams. Since that time several developments have lessened the dichotomy establishing the preconditions for a unified mathematical institutional economics. The work of Zeuthen (1930), Chamberlin (1962), Stackelberg (1934, 1952), and other economists began to bridge the gap between the more formal mathematical models and a richer description of the environment. This was especially true of the work of Chamberlin where emphasis was given to the role of product differentiation and several new and highly relevant variables were introduced into the formal models.

Since the 1930s there has been a considerable growth in literature that is less formally oriented than Cournot (1897), Edgeworth (1925), Hotelling (1929), or von Neumann and Morgenstern (1953), which blends in varying proportions theorizing and analysis, technological facts and an institutional approach. Among these authors are Triffin (1940), Fellner (1949), Brems (1951), Machlup (1952b), Bain (1956), Baumol (1959), Sylos-Labini (1962), Scherer (1970), Sherman (1972), and Williamson (1975).

In the last twenty-five years the growth of the methods of the behavioral sciences, management science, and operations research together with the advent of the computer as a means for manipulating and studying large complex models has brought a fourth approach to the study of the problems of oligopoly. This approach can be loosely called the behavioral or management science theory of the firm. It is typified by the work of March and Simon (1958), Cohen (1960), Clarkson (1962), Cyert and March (1963), Cohen et al. (1964), and Nelson and Winter (1974) and others engaged in simulating complex microeconomic relationships. It is also characterized by the construction of the Carnegie Tech Management

Game, and the work of Shubik (1959b), Marris (1964), and Vickers (1968) on the financial models of the firm.

The behavioral or management science theory of the firm can scarcely be called a theory. It is, however, a new approach leaning far more heavily on the modeling and analysis of detail than previous approaches, introducing more variables than were contained in previous noninstitutional analyses. Behavioral scientists particularly economists are somewhat cavalier in their use of the word theory. Very often a few unproved conjectures are labeled theories. There is presently no single behavioral theory of the firm. Considering uncertainty, managerial incentives and indefinite life of the firm, there is not even a single theory of the profit maximizing firm.

In general there is no single theory of oligopoly. There are a host of partially developed theories based on a mixture of analysis, insight, and to a great extent casual observation. Careful empirical study of oligopoly is expensive and difficult to come by. Few economic theorists have had the time or the resources to avail themselves of large studies. Such studies are usually done only when immediate economic and legal necessity calls for them. Even then the studies are performed to plead a special position. Thus at this time, although the folklore is rich and superficial empiricism plentiful, evidence for theorizing is not readily available.

Mathematical Economics and Oligopoly Theory

Writing in the 1830s Cournot (1838) constructed a formal mathematical model of a market with N competing firms selling the identical product. There was no allowance within the model for the entry of new firms and the firms were assumed to compete via production as their strategic variable. Cournot considered that all firms were attempting to maximize their profits simultaneously, yet independently. He showed that this behavior led to a noncooperative equilibrium that produces a market price higher than the price called for by a purely competitive market. (This concept is defined in detail in Chapters 4 and 5.)

Fifty years later, in a review of Cournot's work, Bertrand (1883) suggested that price, not quantity, should be the independent variable. He formulated a model of duopoly with two firms selling the identical product, using price as the strategic variable. The firms were assumed to have virtually unlimited capacity with respect to the size of the market. Bertrand also assumed that the firms would attempt independently to maximize their profits. The noncooperative equilibrium he obtained with price as the independent variable coincided with the competitive equilibrium.

Edgeworth (1925) writing some years later introduced the additional aspect of limited capacity (steeply rising average costs equivalent to a capacity limit) to the price competition duopoly model. He concluded that there would be no equilibrium whatsoever. To comprehend Edgeworth's work fully it is necessary to understand the nature of demand in an oligopolistic market. (This is discussed in detail in Chapters 5, 7, and 11 and in Appendix A.)

In a stimulating paper, Hotelling (1929) considered two competitive firms selling a differentiated product. The product differentiation was achieved by considering transportation costs and customers strung out along a road between the two firms. He obtained a noncooperative equilibrium that was unequal to the competitive market price.

Subsequent developments were less mathematical in spirit and closer to broad economic theorizing until the advent of the theory of games. The work of von Neumann and Morgenstern (1953) set up the mathematical apparatus for the analysis of conflict, particularly economic conflict or cooperation. For situations involving three or more individuals, von Neumann and Morgenstern concentrated on the development of a cooperative or collusive theory. This was much more in the spirit of the work of Edgeworth (1881), Bowley (1928), and Zeuthen (1930) on bilateral monopoly than in the spirit of Cournot; however, it was far more general and wide-ranging.

Nash (1951) generalized the noncooperative approach. Thus his work is more directly connected with the behavior theory suggested by Cournot, Bertrand, and Edgeworth in his work on duopoly than is that of von Neumann and Morgenstern. Since that time, the application of game theory to the study of oligopoly has been made by Shubik (1959a), Harsanyi (1963), Krelle (1968), Shapley and Shubik (1969b), Selten (1970), Telser (1972), Marschak and Selten (1970), J. W. Friedman (1977), and others. The work has been characterized by the proliferation of a number of solution concepts, that is, the various criteria required to decide when an economic problem is solved may emphasize equity, efficiency bargaining strength, or other factors. Furthermore, entry conditions, bankruptcy, and several other features of the market have been introduced into the analysis. Many of these features are discussed in subsequent chapters.

Along with the considerable growth of work in general equilibrium theory many mathematical economists have, in the past few years, become concerned with the reconciliation of oligopoly theories with general equilibrium theory. This is a sufficiently important new development warranting a separate discussion that appears in Chapter 9.

The first book fully devoted to the application of game theory methods to oligopoly theory after von Neumann and Morgenstern was Shubik (1959b). This volume is to a great extent a development and extension of work presented in Part I of that book. The stress was on noncooperative equilibrium and it was suggested that this solution concept provides a unifying view of much work from Cournot to Chamberlin and beyond. In Part II the importance of the role of information was stressed. Furthermore a dynamic game model was constructed combining the joint considerations of profit maximization with survival. A class of games called *games of economic survival* (Shubik, 1959b; Shubik and Thompson, 1959; Miyasawa, 1962) were considered. Far more concern for the financial structure of the firm was called for in building these games than is usually present in oligopoly models.

Although not represented directly in a book on oligopolistic competition the many writings of Harsanyi have been of considerable influence in the development of game theory methods for the study of oligopolistic competition. Three items of particular interest are the Harsanyi value, his work on games with incomplete information and his most recent work devoted to the selection of a unique equilibrium point for any game played noncooperatively.

The first work (Harsanyi, 1963) was an ingenious extension of the models of Zeuthen (1930) and Nash (1953) to bilateral bargaining. Zeuthen suggested a dynamic model in which concession rates were determined by the relative size of the gains available to each. Harsanyi showed that the Nash bargaining model led to the same outcome as the Zeuthen model and he also generalized the solution for any number of players. His interests then turned to the possibility that not only are many situations best modeled by a game played noncooperatively, but individuals are frequently not completely informed about all of the rules; for example, they may not know each other's preferences. Harsanyi (1967a; 1967b; 1968) approached this possibility by introducing subjective probabilities and Bayesian players capable of updating their probabilities.

Most games, especially those with many stages, when solved for their noncooperative equilibria may easily have many equilibria. Harsanyi (1975) thought it desirable to have some theory and computational method enabling selection of a single equilibrium point as the solution to any game. His tracing procedure offers a means to do this.

The work of Harsanyi must be regarded as both ingenious and normative. There is no evidence, experimental or otherwise, that individuals or organizations behave the way his various works suggest. In particular the whole problem of how individuals update their subjective probabilities still presents more questions than answers. The theories of the decision theorists (Raiffa, 1968) and the evidence of the psychologists (Tversky and Kahneman, 1971) do not fit together. Harsanyi's desires for a theory enabling selection of a single outcome as a solution seem motivated more by aesthetics than empirical considerations. It seems unlikely to some of us that there is or should be a single solution to any oligopoly problem.

Telser (1972) offered a blend of basic game theory and specific empirical analysis applied to the study of oligopolistic competition (see also Aumann, 1973 and Shitovitz, 1973). He has stressed the application of core theory (Shubik, 1959a; Debreu and Scarf, 1963) to market exchange and then to oligopoly. In a further development expectations are introduced and a dynamic model of oligopoly is considered taking into account both cooperative and noncooperative equilibria. His first models are a direct extension of the model of a market with indivisible goods and marginal pairs of traders formulated by Böhm-Bawerk (1923) in his study of a horse market. This has also been treated using game theory by Shapley and Shubik (1972).

Telser's book ends with two chapters devoted to empirical work on estimates of demand conditions and price policies on four products in moderately concentrated markets; and on the determinants of the returns to manufacturing industries. The connection between the empirical work and the theory is somewhat less developed than might be hoped for. Furthermore the linkage between the theory in the book and the nongame theoretic literature such as that of Chamberlin, is undeveloped.

Selten's work has at least two features of considerable importance to oligopoly theory. The first (Selten, 1975) is his development of the concept of a perfect equilibrium point. This provides a restriction on the set of all equilibria by ruling out certain types of threats. In games with low communication and many players, it is reasonable to rule out the use of acts that may be costly to the threatener to carry out if his threat goes unheeded: "If you take another drink I will destroy our only bottle of 50-year-old cognac" may sound brave before the drink has been taken, but after the fact the threatener may wish to reconsider the threat. At perfect equilibrium the threatener would never make a threat that does not remain worth carrying out even in the light of new information.

Another more direct application of game theory to oligopoly theory by Selten (1973) is his construction of a simple model of imperfect competition where four are few and six are many. In essence he studied a Cournot model with symmetric firms and added specific assumptions concerning the institutional possibilities for cooperation. In particular a two-stage game was constructed in which the first stage involves a decision by each firm to participate in a cartel. This is followed by a cartel bargaining stage in which each participant proposes a quota system for the cartel. Finally all firms make their production decisions. At one level the Selten model is ad hoc, yet it nevertheless demonstrates in a clear and striking way how large and small group phenomena can be formally modeled.

The more recent trend in game theory work on oligopoly theory has been toward dynamics. J. W. Friedman (1977) has been concerned with the existence of a special type of noncooperative equilibria in multistage markets and in linking the interpretation of the strategies employed with best reply functions. His concern has been with reaction functions and their stability properties. At a more general level, Sobel (1971, 1973) has been concerned with noncooperative stochastic games and Kirman and Sobel (1974) have applied this work to inventory problems in oligopolistic competition. Marschak and Selten (1974) also offer a reaction function or "convolution" approach to the dynamics of oligopolistic competition.

Economic Theorizing and Oligopoly Theory

Chamberlin's important book represented a considerable step forward in the selection of an adequate model of oligopolistic markets. Yet at the same time as the model of the market was improved in scope, relevance, and realism, the quality and clarity of the analysis was not as high as that found in Cournot or Edgeworth.

The value of Chamberlin's model came from the explicit introduction of product differentiation together with consideration of the entry of new firms. His analysis, however, was not clear. He employed two highly different solution concepts, one for large groups and one for small (in which large and small are not precisely defined). The solution for markets with many competitors proposed by Chamberlin is none other than the noncooperative equilibrium point theory used by Cournot, Bertrand, Edgeworth, and Hotelling, and generalized by Nash. The small group solution, in which interdependence is stressed, is best described as quasi-cooperative with dynamic overtones.

We distinguish truly dynamic solutions from those that have only dynamic overtones using as a criterion the possibility of direct production of a workable computer program from their description. This is merely an operational way of asking whether the dynamics of the moves and countermoves are *explicitly* spelled out and the revenues, earnings, or payoffs over time are described. The works of Stackelberg (1934), Triffin (1940), and Fellner (1949) all fall into the category of quasidynamic in spirit using verbal and diagrammatic demonstrations in practice. Of the three, Triffin's attempt was the most ambitious inasmuch as it attempted to embed monopolistic competition in a general equilibrium framework. Unfortunately no semblance of rigor or plausibility can be gained without more mathematical apparatus than Triffin employed. At the same time the Triffin book added no particularly new empirical insights over the work of Chamberlin.

Both the Fellner and Stackelberg works have been reviewed and digested many times before now. The only point reiterated here is that both stressed a need for dynamics and offered elaborate reaction curve analyses of a relatively ad hoc variety. Fellner presents an easier description of Stackelberg's models than does Stackelberg. But given the proliferation of possible dynamic models one wonders if a verbal analysis which offers neither empirical evidence nor structural detail concerning timing of moves, importance of finance, and so forth, carries us much beyond Chamberlin.

Entry, Exit, and Merger

Entry, exit, and merger are key factors in determining the state of competition. A survey of much of the work on this topic together with a reference list of over 70 items is given elsewhere (Nti and Shubik, 1979). Here only a brief sketch is presented.

Perhaps the key item in the understanding of entry is the dynamics of variation of product or institutional form. Chamberlin with great foresight stressed product variation, but the key qualitative essence was best summarized by Schumpeter describing "competition which commands a decisive cost and quality advantage and which strikes not at the margins of profits and outputs of the existing firms but at their foundations and very lives" (1950, p. 84).

Harrod (1951) offered a critique of Chamberlin's view of entry arguing that it did not take into account long run optimization which could be reflected in short term price policy. Hicks (1954) developed the theme of tradeoff between short and long term profit considerations still further. But it was the publication of Bain's (1956) *Barriers to New Competition* that systematized and discussed many of the major problems concerning entry. Furthermore in this book empirical studies were presented to back up the suggestions concerning "limit pricing" to discourage, slow down, or prevent entry.

Sylos-Labini (1962) attempted to discuss entry in terms of a Walrasian adjustment mechanism in a dynamic context. Working primarily by means of two examples he argued that technological advantage is with large firms. He suggested a minimum return on investment as the entry criterion arbitrarily selected at 5% and suggested that oligopolistic industrial structure weakens the ability of an economy to channel resources efficiently from one sector to another.

Although suggestive, both the mathematical and the institutional economist may have difficulties with the type of theorizing noted above. On the one hand at points which appear to be critical to the mathematical economist the mechanism and the details of timing are not spelled out and on the other hand although the casual real-world references are insightful they lack the explanatory force provided by the relevant detail of a good institutional study.

Shubik (1959a) suggested the concept of "firm-in-being" as a way of both building formal mathematical models of entry and characterizing the class of potential entrants. This concept fits directly into Bain's scheme of classification since the barrier must be measured by a complex of product differentiation, legal, organizational, and financial factors.

Williamson (1965) noted that limit price theory was too deterministic and suggested that the effect of pricing should be interpreted in terms of the probability of market penetration taking into account both timing and number of entrants. Kamien and Schwartz (1975) among others (Gaskins, 1971) have developed mathematical models showing the effect of uncertainty upon limit pricing.

Most of the theoretical work has been directed toward the study of entry, although there has been empirical work on mergers and Ijiri and Simon (1964) have presented a theoretical model of growth. This fact alone suggests that much still remains to be done to understand the phenomena of entry, exit, and growth.

Oligopoly and Organization

In the course of the last twenty-five years there has been a growing recognition that at least for an understanding of oligopolistic competition the model of the firm as an atom or unified decisionmaking unit may not be adequate for most purposes. The amount of detail in the description of the firm and its internal organization or the market and its internal organization were features that traditional oligopoly theory has scarcely been concerned with.

The work of Simon and Bonini (1958) and Ijiri and Simon (1964) provide important examples of how a size distribution can come about under weak conditions involving the interaction of chance and optimization. Balderston and Hoggatt (1962) in an unfortunately little known imaginative simulation of a wholesaler distributor system (based upon the U.S. West Coast lumber industry) present an example of market segmentation, organization, and reorganization based upon relatively slight changes in operating rules and costs. This work is highly suggestive of the value that can be obtained from a blending of simulation and theory.

The text of Vickers (1968) and the more advanced work of Marris (1964) together with the games of economic survival of Shubik (1959b) provide examples of a growing concern with financial structure and details of the internal organization of the firm as an important structural clue to the understanding of behavior.

The difficulty with the more detailed models of the firm and its working is manifested almost immediately. Except in highly special instances it is virtually impossible to study the models analytically. Simulation is called for as illustrated by the work of the Carnegie Tech school and especially by the work of Nelson and Winter (1978) on, for example, the simulation of Schumpeterian competition.

Work such as that of Baumol (1959) and Shubik (1961) was motivated by realization that the size of the organization and the possibility that it is controlled by fiduciaries leads to many different possibilities in the description of the goals of the firm. Baumol suggested that a management may maximize market share subject to an acceptable profit. Shubik found that the top ten priorities among the stated goals of a sample of large U.S. corporations were (1) personnel, (2) duties to society in general, (3) the customer, (4) stockholders, (5) profit, (6) quality, (7) technological progress, (8) supplier relations, (9) corporate growth, and (10) managerial efficiency.

The work of Williamson (1975) represents an important contribution to the modernization of the importance of the belief of John R. Commons that the transaction is the basic unit for microeconomic analysis. In his book on markets and hierarchies Williamson considers in detail the trade-offs between organizational relationships that can best be internalized within the firm—hierarchies, and those that are best handled outside of the firm by markets. It becomes clear as he develops his thesis that there is no neat dichotomy that can suggest immediately when an economic activity should be within the firm or in the market. Transactions costs, timing, uncertainty, legal and societal complications all play their role in determining what may be viable organizations. Information and communication conditions are critical.

Williamson covers the key items in industrial organizations including the employment relationships and the special role of labor; the intermediate products markets and vertical integration; firm size; the multidivisional firm and the conglomerate; and the relationships between market structure and technical and organizational innovation. He suggests that the time is ripe for a new institutional economics emphasizing that (1) received microtheory operates at too high a level of abstraction for most purposes and (2) the transactions approach is at the center of the subject and deserves more attention. We agree with him to a point, but suggest that the time is ripe not merely for a new institutional economics, but for the development of a *mathematical institutional economics* which if anything must be simultaneously more abstract than most of received microtheory, that is, it must deal in abstractions such as flexibility, viability, strategic advantage, information state, and so forth. Yet when applied, the mapping from the abstract concepts to the specific variables must be far more detailed, process oriented, and institutional than is the case with current microeconomic theory. Questions such as what is a transaction or what is a market are key questions. They can be investigated at a high level of abstraction but the application of these investigations requires a considerable level of concretion.

A parallel development in concern for organization of economic units, although from a different perspective, has taken place in France and is illustrated by the writings of Perroux (1975) and others (Wickham, 1966).

OTHER THEORY APPROACHES

There are several important approaches to the study of aspects of oligopolistic behavior that are not detailed in this chapter but merit comment. In particular operations research and management science studies have been applied to a host of specific problems involving inventories, production scheduling, transportation, and product mix decisions. These, although not directly relevant to oligopoly problems, provide the basis for more complex specific models of oligopoly structures. (Some references are noted in Chapter 8; the interested reader is referred to *Management Science* and *The Journal of the Operations Research Society of America* for a sampling.)

There has been a growth in operations research oriented marketing and advertising literature which has been, for the most part, ignored by economic theorists. In the past ten years with the new interest in search and information a new industrial organization has come into being. Despite the close relationship between the two bodies of work there has been little interaction. (References are given in Chapters 10 and 11.)

Another new topic that barely existed twenty years ago is experimental economics. Is there evidence from the laboratory, from the playing of games that favors any of the myriad reaction functions and theories of oligopoly behavior?

Institutional, Econometric, and Legal Studies

The work of Bain on entry encouraged several subsequent empirical investigations. Mann (1966) tested the work of Bain with a larger sample to determine if there was a positive correlation between profit rate, industry concentration, and entry barrier size. Rhodes (1970) and Qualls (1972) obtained conflicting results.

The work of Scherer (1970) provides a wealth of references and reports on empirical studies that serve as evidence for the current health and growth of a better empirically and theoretically oriented study of industrial organization. The size and breadth of Scherer's work would make it virtually impossible to be of equal quality on all fronts. Thus although there is a considerable effort to utilize theory and integrate it with empirical work this is not completely successful. For instance we are told what indices have been suggested to measure monopoly power, but no theoretical justification is given. The book would have been improved by strengthening or omitting the unsatisfactory treatment of game theory.

Despite these few negative comments, Scherer provides an encyclopedic coverage of much of the important work relevant to industrial organization that is needed for problems in the control of industry. The level of common sense and fact makes it a valuable medicine to take by those whose literary diet tends to consist of books on pure theory.

A rare book in its basis in theory and mixed approach at empirical work and experimental economics is that of Sherman (1972). The first two sections deal with experiments involving oligopolistic markets with no entry and concentration on short term considerations; capacity and price choice; and empirical work on oligopolistic behavior.

A Note on Oligopoly and the Law

In the past twenty years there has been a considerable growth in joint work between economists and lawyers. Packer (1963) gave a summary of the state of research in antitrust law up to 1960. Neale (1970) provides a source for current antitrust law. Posner (1969) suggested an approach to oligopoly via the antitrust laws.

With the growth of communication between lawyers and economists, it is of value that they do not lose sight of the great difference in their approaches. There is certainly a tendency by economic theorists to believe that the logic of economic theory (and frequently static economic theory) should be reflected in and match the legal process. The law is what it is and how it is interpreted. Its match with current economic theory will be determined by the sociopolitical climate of the society and by how much economic theorists influence the interpretation of the law. The social process of the law, applied and public policy economics, and economic theorizing have different scopes and purposes. Any attempt to consider them together must recognize these differences.

A Direction for Oligopoly Theory

The thesis of this work is that a fruitful direction for the further development of oligopoly theory calls for a combination of analysis, gaming, simulation and econometrics, and other empirical work. While we believe that we have only made a start in this volume it seems possible to subtract or adhere submodels if we are satisfied with linear and quadratic approximations. These models are amenable to analysis as shown in subsequent chapters, yet they provide a game structure for experimentation and simulation rich enough to merit the measurement of parameters. We present one ad hoc attempt in Chapter 9, not in any belief of its accuracy but to present a model that can be analyzed from the viewpoint of oligopoly theory—gamed, simulated, and parametrically estimated.

Oligopoly theory is of minor importance when compared with the applied problems of industrial organization and the social, political, and legal aspects of the control of industry. Questions concerning tradeoffs between efficiency and equity or decentralization and power involve answers that reflect the sociopolitical biases of a society. Oligopoly theory alone may provide useful but rather narrowly technical insights to the economic aspects of the control of industry. In applying oligopoly theory care must be taken in recognizing where the narrow field of economics ends and the broader field of politicosocioeconomic advocacy begins. This book is confined to the former.

4 | Solutions and Theories of Behavior

WHEN WE SPEAK of a solution to a situation involving conflict or cooperation we are usually referring to a normative plan, that is, a formula or a suggestion that tells the parties how to act, or we are predicting how the parties will act. All of oligopoly theory is prescriptive or descriptive. In some instances, theorizing is intended to change market structure, that is, break up a market, change conditions of entry, or retract special privileges. In other instances there are attempts to legislate against certain forms of market behavior.

Many models offer solutions to oligopolistic competition or collusion. Lawyers have based their suggestions for the control of industry on their interpretations or misinterpretations of some form of economic theorizing, feeling uneasy about the advice but lacking better alternatives. The advent of work in the theory of games, decision theory, and the behavioral theory of the firm together with the methods of gaming and simulation have called attention to the overall problem of the multiplicity of different solution concepts. Broadly speaking there are four major solution-concept categories of relevance to the study of market behavior. They are (1) cooperative solutions, (2) noncooperative solutions, (3) mechanistic solutions, and (4) dynamic or quasidynamic solutions. The contrast between the fourth category and the other three may at first seem strange; however, the first three are well defined in a static context, but their distinction blurs in a dynamic situation. The fourth category contains behavioral models, encompasses the possibility of quasicooperation, and is the natural category for theories devoted to building artificial players, that is, computer programs that simulate the behavior of firms in oligopolistic competition.

The Strategic Model of a Game

An n-person game in strategic form consists of n individuals called *players;* these players can be firms, armies, persons, or institutions. Each player is assumed to have some method of evaluating the personal worth of any outcome. Each player has preferences. Furthermore, each player i has available a set of alternatives or strategies S_i. In particular, we use the lower case notation s_i to denote a specific strategy belonging to the set of strategies S_i of player i.

It is assumed that *all* of the *rules of the game* are known to all players. This is a *very* strong assumption, but it aids in formulating problems amenable to analysis and can be relaxed later. A simple example serves to illustrate this point: when playing chess it is reasonable to assume that each player prefers a win to a draw and a draw to a loss. Furthermore, each player knows all of the rules. This implies that all resources are known. In the case of chess the player knows the pieces, how many he should have, and how they move. He knows the initial disposition of the pieces, what a chess board looks like and how to set it up. He knows all the rules of capture. In short the player knows how to play chess.

There are many poker stories in which the stranger walking into town, gets into a poker game, and at a critical moment (when losing) discovers that there is a special house rule that is working against him. If he chooses to dislike the rule he had better be faster with his six-shooter than his hosts. This is an example of an incompletely defined game and is to a great extent a better model of business enterprise than chess. However, to begin we restrict ourselves to the former rather than the latter model.

If each player selects a strategy in the game, the resultant set of strategies $(s_1, s_2, \ldots, s_n)$ determine an outcome. Since each player has a method for evaluating all outcomes we associate a payoff function with each player, which describes his evaluation for all outcomes. In particular, player i has a payoff function $P_i(s_1, s_2, \ldots, s_n)$. There are n payoff functions and each payoff function is a function of n variables.

A strategy can be a very complicated plan such as a complete plan for playing chess, or a total long-range plan replete with decisions covering all contingencies for developing a market. It is easy to see that in most situations individuals or institutions rarely use strategies. They usually operate with a more-or-less plan if they have any plan at all. The plan may cover a few moves and then it is revised. Attempts to devise computer programs for playing chess work on such a basis. There are some simple one-shot games, however, for which it makes sense to talk about strategies and they are of considerable use for purposes of illustration.

A simple example of a two-person game in which each player has a set of strategies consisting of only two alternatives is of great value in illustrating many of the aspects of competition and cooperation. Table 4.1 shows two 2 × 2 matrices (a and b) which can be written together to save space as is shown in part (c). For ease of exposition the players are called A and B. The numbers on the left-hand side of each matrix represent the strategies of Player A; he can choose 1 or 2. The numbers on the top of each matrix represent the strategies for Player B; he also can choose 1 or 2. The numbers in Table 4.1(a) are the payoffs associated with the four outcomes (1, 1), (1, 2), (2, 1) and (2, 2). Thus the matrices 1a and 1b represent the payoff functions $P_1(s_1, s_2)$ and $P_2(s_2, s_2)$ of Players A and B where s_1 and s_2 can take on the values 1 or 2. The third matrix, part (c), has two values in each box. The first value is the payoff to the first player; the second is the payoff to the second player.

Table 4.1 is a classical game known as the Prisoner's Dilemma. This game has been extensively studied and many experiments have been run with it (Rapoport and Chammah, 1965). Rapoport and Guyer (1966) calculated that there are 78 structurally different 2 × 2 matrix games with ordinal payoffs.

Prior to discussing solutions in a formal manner it is instructive to examine this game in the light of our concepts of individual rationality. Consider Player A. He might reason as follows:

> Suppose that B elects to choose his first strategy. If I use my first strategy then I would obtain 5. I would obtain 10 if I were to use my second strategy. I prefer 10 over 5.
>
> Suppose that B elects to choose his second strategy. If I use my first strategy then I would obtain −10. I would obtain 0 if I were to use my second strategy. I prefer 0 over −10.
>
> Whatever he does I am better off with my second strategy. Hence I should use it.

We can apply the same type of reasoning to Player B and come out with the same conclusion based upon individual "rationality." Each should use his second strategy. This yields a payoff of 0 to each. Had they both employed their first strategies the payoff would have been 5 to each.

Table 4.1 Simple example of a two-person game.

					B's STRATEGIES				
A's STRATEGIES		1	2		1	2		1	2
	1	5	−10	1	5	10	1	5, 5	−10, 10
	2	10	0	2	−10	0	2	10, −10	0, 0
		(a)			(b)			(c)	

In this simple example it can be seen that individual rationality employed without cooperation does not necessarily lead to an outcome that can be described as societally rational, efficient, or Pareto optimal. Both players could improve their payoffs above the level of 0 to each.

Cooperative Solutions

Oligopoly theory generally is discussed from society's view of the behavior of a group of firms. Thus, actions that the firms might term cooperative are considered collusive by other members of society. It is important to remember that our analysis is almost always partial. Advocates for control of industry are concerned with the dichotomy between large agglomerations of industrial power and the relatively powerless consumers. Their investigations are directed specifically toward an understanding of the complex of interactions between firms and their customers who are also often their owners and employees.

We make no pretense in this work at answering many of the vital welfare questions that can be posed concerning the complex of relationships among firms, customers, stockholders, employees, and others. We concentrate on the problems of competition among few large firms in which competition can best be described as noncooperative or quasicooperative. From our viewpoint the customer is replaced by a demand relation. If we wish to bring the customer into the calculations explicitly, he will be a strategic dummy (Shapley and Shubik, 1972), that is, an individual who has preferences that count but who has little, if any, power in the immediate economic arena. This, of course, does not imply that the consumer or stockholder does not exercise power by voting or other means. It does mean that he has no individual market power when buying a pack of cigarettes.

Ignoring the individual consumer as a strategic power in the market, we concentrate instead on the behavior of the firms as a group. Cooperative solutions, like pure monopoly, represent an extreme form of static analysis. All players are assumed to optimize jointly. Even for the study of cartel arrangements and collusive agreements among firms the cooperative solutions are not satisfactory. The very essence of cooperative arrangements in the market depends upon the *dynamics* of the situation and almost always a static model of the process is inadequate.

Although we suggest that cooperative solutions[1] are not particularly useful for the study of oligopolistic collusion, there are at least two solu-

1. Since our interest is oligopolistic competition, the other major cooperative solutions are not discussed. A full description of these solutions is given elsewhere (Shapley and Shubik, 1974). The major solutions are the core, the value, the bargaining set, and the stable set; other solutions include the kernel, the nucleolus, and the β-core.

tions that provide valuable benchmarks. They are the Pareto optimal set and the point of joint maximization. These serve a purpose in the construction of measures to decide upon the degree of collusion feasible in a market.

The Pareto optimal set is that set of outcomes such that the payoff to no individual can be improved without diminishing the payoff to at least one other individual. This is an efficient solution in the sense that no resources are wasted by the group. If there were, it would be possible to improve the payoff to at least one individual at no cost to the others.

The joint maximum is a special point in the Pareto optimal set. *If* it were possible to directly compare individual welfare using a common numerical scale, it would be meaningful to talk about aggregate welfare. While it is not always reasonable to compare consumers' welfare, the profits of corporations in the same market can be compared for many purposes.

Both the Pareto optimal set and the joint maximum are illustrated in Figure 4.1 using the example given in Table 4.1(c). The horizontal axis denotes the payoff to the first player and the vertical axis denotes the payoff to the second. Hence the point (10, −10) illustrates the payoff resulting from the strategy pair (2, 1), that is, the lower left-hand of the matrix (c) in Table 4.1.

The joint maximum occurs at the point (5, 5) where the overall gain to be had under complete collusion or cooperation is 5 + 5 = 10. This outcome, however, is not the only one in the Pareto optimal set. Both

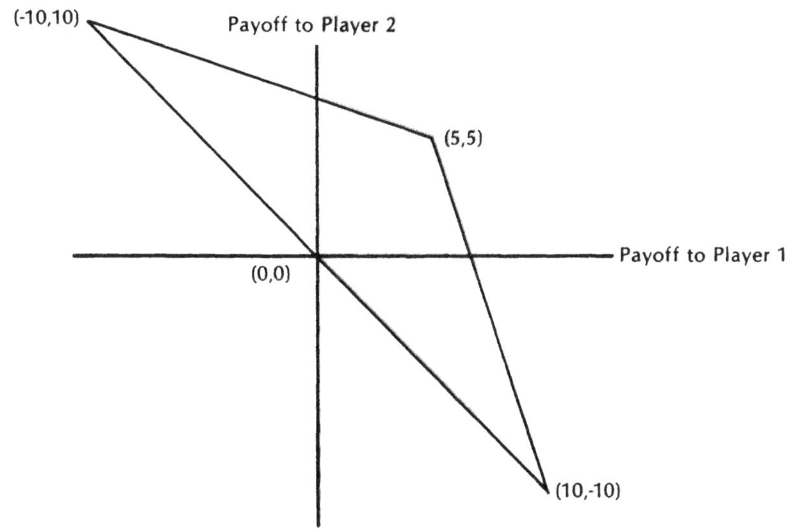

Figure 4.1. The prisoner's dilemma.

(10, −10) and (−10, 10) are also in this set. In the first instance there is no way of improving the lot of the second player without taking from the first; and in the second case, vice versa.

The joint maximum may be expressed mathematically as

$$\max_{s_1 \in S_1} \max_{s_2 \in S_2} [P_1(s_1, s_2) + P_2(s_1, s_2)],$$

where S_1 and S_2 represent the set of alternative actions available to players 1 and 2 respectively; s_1 and s_2 refer to a specific action. The mathematical description and analysis of the Pareto optimal set is given in Chapter 5 where we examine the difference between the set obtained from considering only the welfare of the firms and the set where the welfare of the firms and the customers are both included.

Contract and Cooperation

A key element in cooperative theory is the implicit assumption that contract is not violated. Formal cooperative game theory avoids the problem posed by the dynamics of trust and negotiation. All rules are given and obeyed; all contracts honored. This enables study of the combinations of cooperation in a static context abstracting from dynamics, law, and the problems involved in the evolution of trust and communication.

An adequate theory of oligopolistic cooperation must include the *process* of cooperation. Multistage games are the analytical tools that may be used to examine noncooperative equilibria reflecting complex information conditions, signals, and threats. Although the solution suggested here can be described as noncooperative, the possibilities of communication and threat may lead to outcomes that may be viewed as quasicooperative with cooperation enforced by a self-policing system. The connection between law and economics is clearest when we contemplate the dynamics of collusion. The law provides many of the rules in the dynamics of trust and contract formation.

Collusive or cooperative oligopolistic behavior is omitted from discussions in this work not because it is deemed unimportant. On the contrary it is not presented in any detail because the static theory of cooperative games cannot adequately answer many questions of interest. The treatment given in this work serves as a preliminary to dynamic theory.

In a practical sense, detailed institutional analytical studies taking into account both the law and specific details of industrial organization will provide a better basis for the construction of a dynamic economic theory of oligopolistic behavior than will any pure a priori reasoning. The details on time lags and the special weapons of cooperation or competition be they technology, special legal, social, or other economic variables provide a structure for the institutions which are the carriers of process.

Noncooperative Solutions

Cooperative solutions have in common the idea that all players (in this case firms) play efficiently in the sense that they will not jointly forego possible gain. This was succinctly summarized in the rather general cooperative solution concept of Pareto optimality. When we deal with noncooperative solutions, the assumption of optimality is not made. This does not mean that a noncooperative solution may not have as an outcome a Pareto optimal distribution. Pareto optimality could easily arise from the structure of the market. However, it is not a necessary condition for noncooperative behavior.

The most famous but by no means unique noncooperative solution concept is that of the noncooperative equilibrium point. This was described in its most general form by Nash (1951). However, it also (in more specialized manifestations) was described by Cournot and Edgeworth, and serves as the basis for the equilibrium suggested by Chamberlin in his study of monopolistic competition. Let us represent the set of alternatives facing the first player by S_1 and the set of alternatives facing a second player by S_2. A specific alternative to player i is denoted by s_i. A pair of alternatives (s_1^*, s_2^*) form a noncooperative equilibrium point if they satisfy the following conditions:

$$\max_{s_1 \in S_1} P_1(s_1, s_2^*) \Rightarrow s_1 = s_1^*,$$
$$\max_{s_1 \in S_2} P_2(s_1^*, s_2) \Rightarrow s_2 = s_2^*.$$

(The arrow stands for the word implies.) We see that these two conditions call for a type of circular stability. If the first player is aware that the second player is going to select s_2^*, the first player will select $s_1 = s_1^*$ when he maximizes his own payoff. If the second player is aware that the first is going to select his alternative s_1^*, second player will select $s_2 = s_2^*$ when he maximizes his payoff.

Individual maximizing behavior may, of course, lead to societal disaster as we have already seen in the simple game given in Table 4.1(c). The noncooperative equilibrium in this case calls for the strategy pair (2, 2) to be played which yields the outcome (0, 0) as payoffs to the players. This is illustrated in Figure 4.1 and is a highly undesirable point located a considerable distance from the Pareto optimal set. As we shall see with some examples later, this is not necessarily, but often is, the case.

With experimental games more so than in actual markets, participants convert a situation that is quite clearly nonzero sum, that is, a situation in which there are the possibilities for mutual gain, into a contest involving strict opposition. This can be done by trying to maximize the difference between the scores of the two participants. The behavior can be described as

$$\max_{s_1 \in S_1} \min_{s_2 \in S_2} (P_1(s_1, s_2) - P_2(s_1, s_2)),$$

which converts a nonconstant sum game into a constant sum game. The meaning of this form of behavior is clear. This type of calculation is often made in military circles when investigating damage exchange ratios. In some sense we may regard the striving of each competitor to maximize the difference in payoffs as the Alice-through-the-looking-glass companion of the cooperative solution that calls for joint maximization. In the case of joint maximization the individuals were concerned about their welfare. In the case of the maxmin-the-difference-in-payoffs solution, the individuals are concerned about each other's illfare. In particular, we determine the joint maximum by finding the point at which for one player to increase the payoff to the other player by one dollar the cost to the first player is precisely one dollar. When we are looking for the maxmin solution we investigate the damage exchange rate selecting the point at which for one player to inflict one dollar's worth of damage on the other he has to expend one dollar.

It can be argued that this particular solution concept is related to certain cooperative solutions in the sense that prior to carrying out certain negotiations firms may wish to threaten each other to secure a more favorable initial position from which to start any bargaining procedure. The maxmin of the difference in payoffs serves to set a zero point in the sense that one firm can say to the other, "Look, if we go to war, we can measure the relative damages we can do each other and can calculate the relative costs."

When there are more than two firms the difference game is not adequately defined. However, if the firms are of the same size, that is, if the game is symmetric, there is a natural extension of the maxmin-the-difference solution, and that is beat-the-average. The idea behind the beat-the-average solution is that each firm looks at the rest of the market in aggregate and asks itself, "Am I doing better than the average?" It is of interest to note that this is often the type of reasoning used by financial analysts to select the best firm in an industry. Comparative analysis when information is hard to come by tends to stress relative standing and to increase the competitive view of the process.

Suppose there are n firms. We may define the beat-the-average solution as follows:

$$\max_{s_i \in S_i} \left\{ P_i(s_1, s_2, \ldots, s_n) - \frac{1}{n-1} \sum_{j \neq i} P_j(s_1, s_2, \ldots, s_n) \right\}, \qquad \text{for } i = 1, 2, \ldots, n.$$

It is easy to see that when $n = 2$ the beat-the-average solution coincides with the maxmin-the-difference solution.

When a market is highly nonsymmetric the beat-the-average solution does not have much meaning. It must be further modified to catch the spirit of extreme competition, yet at the same time reflect the structure of the market or other situation. One way of doing this is to consider the struggle as a game-of-status (Shubik, 1971). In a game-of-status, the numerical outcomes per se are not important. The only feature that counts is the position or order of the payoffs. All individuals are striving to be first. The potlatch takes precedence over production in a society such as this. There can be many millionaires. However, there can only be one richest man. To some it may be preferable to be the richest individual in a society of paupers than to be mildly rich in a society of millionaires. We do not pursue the game-of-status solution further at this point but turn to two other generalizations that are less sociological but more relevant to the marketplace.

A natural extension of the beat-the-average solution in the marketplace is maximize-profit-share. This can be described as

$$\max_{s_i \in S_i} \frac{P_i(s_1, \ldots, s_n)}{\sum_{j=1}^{n} P_j(s_1, \ldots, s_n)},$$

where we assign the value of $1/n$ to $0/0$. It can be seen that if all of the firms are symmetric the goal of maximizing profit share is equivalent to beat-the-average.

Another noncooperative possibility is that each firm strives to maximize its share of sales, by which we mean gross revenues. This is not necessarily the same as trying to maximize the share of profits in the market, but very often the two are closely correlated. This solution has been noted in Chapter 2.

One further generalization in the economics of illfare is the threat set or the Pareto minimal set. This has the same Alice-through-the-looking-glass interpretation in relation to the Pareto optimal set as the maxmin-the-difference solution has to the joint maximum solution. The Pareto minimum set consists of those outcomes such that it would not be possible to hurt the competitor more without hurting oneself further. The threat curve is illustrated along with the Pareto optimal set in Chapter 5 when we turn to our first economic models.

There remains a whole set of possibilities of which we note three psychologically interesting solutions to the two-person game. They are

$$\min_{s_1 \in S_1} P_2(s_1, s_2), \quad \min_{s_2 \in S_2} P_1(s_1, s_2).$$

The expressions above represent the most extremely paranoid type of behavior possible. Each player is concerned with doing as much damage to the other as possible. To make this solution completely well defined we must add the boundary condition that each player is constrained to avoid destroying himself in his attempt to render the largest amount of damage to his competitor.

The next solution is

$$\max_{s_1 \in S_1} \min_{s_2 \in S_2} P_1(s_1, s_2),$$
$$\min_{s_2 \in S_2} P_1(s_1, s_2) \quad \text{subject to} \quad P_2(s_1, s_2) \geq 0.$$

In this instance the second player is out to damage the first. However, the first player is aware of it and plays conservatively to minimize the amount of damage that is done to him.

The final solution is given by

$$\max_{s_1 \in S_1} \min_{s_2 \in S_2} P_1(s_1, s_2), \quad \max_{s_2 \in S_2} \min_{s_1 \in S_1} P_2(s_1, s_2).$$

In this particular case, we see that both players are cautious in the extreme. Each assumes that the other is out to do him as much damage as possible. In some cases, such behavior calling for excessive conservatism on all parts may lead to the possibility that "the meek shall inherit the earth." This is not a general result but will depend on specific market structure. When there are more than two players it is evident that there will be many combinations of different behavioral types leading to different solutions.

A Sociopsychological Solution Concept

One further solution provides a parametric measure of the degree of cooperation present in a market. Let us denote by Π_i the social payoff to the ith player, defined as a weighted average of the payoffs to all other players and to each individual i:

$$\Pi_i = P_i + \Theta_i \sum_{j \neq i}^{n} P_j,$$

where the parameter θ_i is a measure of the ith player's cooperative attitude to his fellow players.[2]

The relationship of this cooperative attitude measure to some of the other solutions may be seen immediately. For the parameter setting $\theta_i = 1$, where $i = 1, 2, \ldots, n$, the individual maximization of Π_i be-

2. The parameter θ_i was first suggested by Edgeworth (1881) who called it the coefficient of sympathy (p. 53).

comes equivalent to joint maximization. Each individual is noncooperatively motivated to maximize the joint sum of the payoffs.

When $\theta_i = 0$ for $i = 1, 2, \ldots, n$, each individual is only concerned with his own payoff and the maximization of Π_i is the same as the noncooperative equilibrium described previously. Setting $\theta_i = -1/(n - 1)$ yields the beat-the-average solution.

We have included this solution concept as a subset of the noncooperative solutions because the solution calls for joint maximality when the $\theta_i = 1$ through purely independent action. The individuals have been previously socialized to the extent that they are implicitly concerned with the welfare of the others. The meaning of implicitly concerned is that the welfare of the others is directly considered in the payoff functions of each individual and is *not* a construct outside these payoff functions.

When we speak of joint maximality we are usually referring to the situation in which individual payoff functions are given and individuals collectively make explicit agreements regarding joint behavior. It should be stressed here that the difference in interpretation of the same mathematical operations is critical. We are dealing with long-run possibilities concerning the nature of preferences in society. The joint maximizers are presumed to really love their fellowman when $\theta_i = 1$; when $\theta_i = 0$, they may be willing to cooperate because each can see the personal gain from cooperation. Another way of phrasing this is that in one situation individuals can see explicitly the possibility of gains from trade and cooperation among themselves. In the other instance individuals have concern for their fellow players and act accordingly without necessarily having to debate with them, coordinate with them, or expect an immediate "explicit return from trade."

A Mechanistic Solution

The competitive price system in a market may be regarded as a mechanistic process for obtaining decentralization and efficiency under special circumstances. Although it is demonstrated in Chapter 6 and elsewhere (Shapley and Shubik, 1969a) that the price system may be regarded as a limiting form of noncooperative or other behavior, it is generally discussed in mechanistic form. In other words individuals are assumed to possess complete knowledge about their preferences and costs (or at least complete knowledge to the limit of probability distributions); to believe that they face an impersonal mechanism and have no influence on the market; to be constrained to maximize their welfare or profit. Individuals are not involved in a game of strategy and in fact have no free will in the sense that they are not even permitted to err in this formulation of the functioning of the price system. We could just as well replace the individual entrepreneur by a simple mechanism that shows how to obtain the maximum of a function involving essentially only one variable.

Not all games have an associated price system; however, all situations for which there is a price system can be regarded as a special or limiting form of some game. This being the case, it is worth noting the price system as a solution concept, when discussing forms of solution to competition among firms. Considered in this light, it should come as no surprise that a price system is of equal if not greater importance to a socialist economy that it is to a capitalist economy. A price system involves decentralization and efficiency in which individual profit centers or entrepreneurs can be treated as nonstrategic mechanisms. It is as though they are well behaved branch commissars whose interests to innovate or find other methods for gaining strategic control over the environment do not exist.

Market Form and Behavior: Simple Examples

Types of Competition in Different Market Forms

The market form of an industry is a function of the number and resources of competitors and possible entrants, the physical nature of the market, the legal structure of the country, and the information state of the competitors. These variables limit the alternatives that are legally and technologically feasible for the players. The behavior of the players is determined by the intents of the firms and the market form.

In the short run, market form is only slightly affected by the behavior of the firms. In the long run, however, the behavior of the firms may actually change the market form. For instance, after a period of competition all competitors except one may be eliminated. Or an informal coalition agreement may become a law and hence become part of the game rules. The cartelization laws in Germany and the codes of the early Roosevelt era provide two such examples. These long-run effects must be taken into account when describing the long-run evolution of a market, but in most legal cases and in much of economics the inclination to examine the broad and majestic sweep of the long run must be tempered by carrying out an analysis of the short run. Questions such as whether firms A and B are colluding must be answered.

In certain market forms, short-run competitive behavior of firms in an industry takes on a character that is impossible for an external observer such as a government agency to diagnose as completely cooperative, noncooperative, or cut-throat. For example, consider a simplified duopoly with the following payoff matrix:

	B's strategies	
A's strategies	#1	#2
#1	(2, 1)	(2, 4)
#2	(6, 1)	(6, 4)

The market form is reflected in the various payoffs that may result from the strategies chosen by the two firms. These strategies can be interpreted as production decisions. Suppose that no firm has control over price and that advertising and other forms of product differentiation do not matter. If the first player chooses the first strategy he will obtain 2 no matter what strategy his competitor chooses. Similarly, the second player's payoff to a certain strategy is separate from his competitor's. If such a market actually exists, all forms of behavior previously described are indistinguishable! As the fortunes of the firms are not strategically interlinked it becomes meaningless to talk about joint maximization, cut-throat behavior, and so forth. The economic structure forces each player to follow a certain course of action regardless of his attitude toward fighting or colluding. Different intents are unidentifiable. In all cases both players select their second strategies and obtain 6 and 4 respectively.

The following payoff matrix provides an example of an oligopolistic market form in which only a few firms compete: Suppose that there are two firms whose total capacity is in excess of the total demand for their product. Furthermore, suppose that the market is insensitive to advertising and that there are no major opportunities to innovate or improve product (an extractive industry such as coal might provide an example). The strategies of the two firms are either to accept implicitly or explicitly some share of a market in which both take care not to flood, dump, or undercut, but to react in some manner against any disturbance; or to undercut, flood the market, or otherwise engage in unrestrained competition.

	B's strategies	
A's strategies	#1	#2
#1	(5, 5)	(0, 2)
#2	(2, 0)	(−3, −3)

The payoffs in the matrix indicate that if the firms coexist peacefully they can each obtain a revenue of 5 from the market; if either takes another action they both fare less well; and if both indulge in strategies that have been traditionally called competitive, they will both lose heavily.

Joint maximization and noncooperative equilibrium call for the players to use their first strategies and both give the same result, a yield of 5 to each. We cannot distinguish between two behaviors in the marketplace if this is the market form.

If the firms were intent upon damaging each other they would use their second strategies. Hence it is possible to distinguish cut-throat competition (implying maxmin $(P_1 - P_2)$ or beat-the-average) from other behavior patterns.

The following matrix illustrates the competitive feature of a dynamic oligopolistic market in which variables such as innovation and advertising play a role:

	B'S STRATEGIES	
A'S STRATEGIES	#1	#2
#1	(10, 10)	(6, 12)
#2	(12, 6)	(7, 7)

The numbers in this matrix must be interpreted as expected values rather than returns that will definitely result in an outcome of two strategies. The first strategy may be a plan to restrain outlays on innovation, research, and selling effort, combined with a recognition of a comfortable market share providing one's opponent does likewise. If the opponent does not conform to the desired action pattern, some type of countermeasure will be put into force. The second strategy calls for a pattern of considerable spending in innovation and selling regardless of the actions of the competitor.

The interpretation of the four possible outcomes is as follows: If both firms are "peaceful," they can each make a profit of 10. However, there is the possibility of an extra profit, some of which may be obtained by getting a larger market share, for a firm that is willing to lead in sales or innovation. The firm may not necessarily get this profit if its development or sales effort fails, but at least it increases its profit expectation. If both firms are committed to a program of innovation and sales effort from the start, their expected profit is 7 each.

The behavior characterizing joint maximization indicates that the firms should each employ their first strategies. The other intents will lead to the firms using their second strategies. Note that in the first two market forms it was impossible to distinguish the results of a jointly maximal behavior from noncooperative equilibrium; here we are able to do so.

The feature distinguishing many parlor games from most economic situations is that in the former the interests of the competitors are diametrically opposed, whereas in the latter this is usually not so. The amount of money won by one player in a two-man poker game equals the amount lost by his opponent. If one player wins a chess game his opponent loses it. This clear-cut opposition is seldom encountered in economic affairs. An example of where such a situation may be a good approximation to reality is in the selection of advertising programs. The payoffs in the following matrix could arise if the firms were in a market in which they compete for each other's customers by using fixed budgets in different ways. This matrix portrays a constant sum game. The sum of the payoffs resulting from the employment of any pair of strategies is constant.

	B'S STRATEGIES	
A'S STRATEGIES	#1	#2
#1	(2, 1)	(3, 0)
#2	(1, 2)	(4, −1)

In this market situation noncooperative and cut-throat behaviors result in the same action. Both players use their first strategies. As all outcomes yield the same sum for the joint payoffs, jointly maximal behavior does not call for any particular course of action to be followed. All pairs of strategies are jointly maximal.

The heavy industry oligopolistic form described in the following example is closely akin to the previous one, although it includes some added market specifications.

	B'S STRATEGIES		
A'S STRATEGIES	#1	#2	#3
#1	(5, 5)	(1, 6)	(−5, 2)
#2	(6, 1)	(2, 2)	(−4, −2)
#3	(2, −5)	(−2, −4)	(−3, −3)

Suppose the competitors are in an industry in which they have considerable capital investment and plant costs. Market share can possibly be expanded by innovation and sales effort, but retaliation to price-cutting is easy and can quickly result in the firms selling below cost. The three strategies of the players can be interpreted as follows: (1) A call for a peaceful market division with limited expenditures on innovation and sales and no price activity if it is observed that the competitor is doing likewise. If the competitor is observed to be active in innovation, then some type of counterpolicy involving innovation is followed. If the competitor is active in price competition, then a counterpolicy involving price is followed. (2) A plan for innovation, possibly with contingent behavior in case the competitor follows different policies, but its essential feature is to follow some innovation policy. (3) A price-cutting policy.

If the firms are jointly maximizing, they will each pick their first strategies. If they follow the noncooperative behavior, they may reach one of two different equilibrium positions. These are obtained if both players use their second or if both use their third strategies. If the firms follow a cut-throat behavior, they will each pick their third strategies. In this market it is always possible to distinguish between completely cooperative behavior and the others by observing the outcome. It is sometimes possible to distinguish between noncooperative competition and the rest depending upon which strategies are employed.

Market Form or Market Behavior?

In some of the above market forms, the results of different types of behavior are indistinguishable. Concepts such as implicit collusion, competition, the intensity of competition, collusion, and cut-throat competition cannot be given meaning independent of market form. Different market behaviors may give the same market results in some market forms. The same behavior may produce different phenomena in different markets. For instance, the benevolent intent of one market gardener to provide low-cost vegetables may lessen his short and long-run profits considerably but have little effect on the profits of his competitors. The benevolent intent of General Motors to provide low-cost automobiles could result in the elimination of some competitors.

The problem of social control of competition involves consideration of ethical and sociopolitical factors, but it also entails an understanding of the logical problems that can arise in attempts to define, understand, and control economic power. The few simple examples provided here merely serve to illustrate how varieties of structure can make it extremely difficult and in some cases impossible to deduce intent from behavior. It remains now to build market models with considerable economic content and to apply our different solution concepts to them.

5 | A Symmetric Market Model and Game

IT IS APPROPRIATE at this point to begin discussion of the formal structure and analysis of a specific model of an oligopolistic market. The mathematical model of this market is specified and investigated. Prior to treatment of the complete model with product differentiation the special case in which each firm sells the identical product is examined. Although it is a special case in the sense that it can be obtained from the more general model by setting a parameter equal to zero, it is worth considering because of its historic connection to the work of Cournot (1838) and because of its use in exposition.

Duopoly with Identical Products: Production Strategies

Suppose that two firms are selling an identical product, no immediate potential entrants need to be considered, and the firms have fixed average costs and a capacity limit. The last two criteria specialize the assumption of the U-shaped average cost curve. However, from both the viewpoint of measurement and computation they are not unreasonable. Cournot (1838), Stackelberg (1934), and others, especially Wald (1951), have discussed more general conditions for the existence of certain types of solution to competition in a duopolistic market. J. W. Friedman (1977) provides a clear and careful exposition of sufficient conditions for the existence of a noncooperative equilibrium in the class of Cournot oligopoly models.

Here we assume that the demand for the product is linear and is given by

$$q = \alpha - \beta p, \tag{5.1}$$

where q is quantity demanded and p is price of the product.

We assume that the firms have equal average costs, c. Thus $C_i(q_i) = cq_i$ for $i = 1, 2$ represents total costs; each firm has a capacity of k_1 and k_2 respectively. Detailed examples with nonsymmetric costs and capacities are given elsewhere (Shubik, 1959b; Levitan and Shubik, 1972) and some examples are discussed in Chapter 9.

Cournot Model

The large firm in competition with few others has the choice of using price, production, or price and production as its strategic variables. All combinations are possible and the importance of each varies with the institutional details of the market.

If the firms choose to compete through quantity (this might be the situation when an agricultural product must be planted well in advance of the market), they offer their product to the market and there is some process by which a price clears the market. If the firms name price, we must determine how the market is allocated between them. They could name different prices if they were soliciting business on a produce to order or a sealed bid basis.

In some instances a firm is required to both name a price and select a level of production. It runs the dangers of being caught with excessive inventories or being out of stock at a critical period.

In general the best representation of the strategies employed by a large part of manufacturing industry involves treating both price and production as independent strategic variables. Depending upon the technical details of the industry the flexibility of each of these variables will vary considerably. In particular, it is relatively easy to change the price of cigarettes; it is less easy to issue changes to prices listed in large catalogues. It is also easier for a manufacturer to change the price of his product if he controls his distribution system than if he had to deal with independents. It is difficult to change the quantity of production for agricultural products with fixed growing seasons and often large job shops may be relatively inflexible owing to bottlenecks, capacity constraints, or supply difficulties.

The simplest model to treat at the outset of an investigation of market structure and behavior is that in which quantity is taken as the independent variable. Let the two firms each select a level of production q_1 and q_2. The total supply offered to the market is denoted by $q = q_1 + q_2$. From

Eq. (5.1) the market price will be given by

$$p = \frac{\alpha - q}{\beta}, \qquad \text{for} \quad q_1 \leq k_1 \quad \text{and} \quad q_2 \leq k_2. \tag{5.2}$$

The revenue functions or payoffs may be expressed as

$$P_i(q_1, q_2) = q_i\left(\frac{\alpha - q}{\beta}\right) - cq_i$$

$$\text{for} \quad i = 1, 2, \quad q_1 \leq k_1 \quad \text{and} \quad q_2 \leq k_2. \tag{5.3}$$

On the assumption that both firms have sufficient capacity[1] the *noncooperative equilibrium*[2] must satisfy the first-order conditions,

$$\frac{\partial P_i}{\partial q_i} = 0 \qquad \text{for} \quad i = 1, 2, \tag{5.4}$$

which yield

$$\frac{\alpha - 2q_1 - q_2}{\beta} - c = 0 \quad \text{and} \quad \frac{\alpha - q_1 - 2q_2}{\beta} - c = 0, \tag{5.5}$$

or

$$q_1 = q_2 = \frac{\alpha - \beta c}{3}. \tag{5.6}$$

It is easy to check that $\partial^2 P_i / \partial q_i^2 < 0$; hence Eq. (5.6) is a result of independent maximization.

The profits made by each firm (excluding fixed costs) are

$$P_1 = P_2 = \frac{1}{9\beta}\{\alpha - \beta c\}^2. \tag{5.7}$$

Breakeven Solution

Suppose that each firm has a fixed cost K, then the profit to each is given by

$$P_i(q_1, q_2) = q_i\left(\frac{\alpha - q}{\beta}\right) - cq_i - K. \tag{5.8}$$

If K is large enough it can exceed the size of profits shown in Eq. (5.7); hence, both firms would yield losses at the previous noncooperative

1. When this is not the case, the boundary solutions must be calculated. See Chapter 8.

2. This is not the only type of noncooperative equilibrium. Historical strategies based on an ability to communicate threats give rise to other noncooperative equilibria related to the well known "kinked oligopoly demand" analyses (Sweezy, 1939). These are discussed in Chapter 6.

equilibrium outputs. This cannot be stable over many periods and we would expect at least one firm to exit or eventually liquidate.

An alternative to accepting losses or exiting is to each restrict production so that the firms break even and cover fixed costs. This calls for

$$q_i\left(\frac{\alpha - 2q_i}{\beta}\right) - cq_i = K$$

or

$$q_i = \frac{(\alpha - \beta c) \pm \sqrt{(\alpha - \beta c)^2 - 8K\beta}}{4} \tag{5.9}$$

which will be positive for $8K\beta \leq (\alpha - \beta c)^2$. Such behavior requires coordination without cooperation. The breakeven solution is not stable. When $K = 0$ this solution coincides with the efficient or competitive solution, as can be seen in the following section.

In a duopolistic market the breakeven level is in general not unique. There may be two breakeven points with the first at B_1, with price higher and production lower than at joint maximum (where $q_1 = q_2 = A = [(\alpha - \beta c)/4]$). The second breakeven point has a larger production as is shown at B_2. This breakeven level is illustrated in Figure 5.1.[3]

The difference between the duopolistic breakeven chart in Figure 5.1 and the competitive breakeven chart in Figure 2.7 should be noted. In the latter chart, any production above the point B means more profits; here positive profits are restricted between B_1 and B_2.

3. In the actual play of a duopoly game, a reversal in price and the production rates after a price-cutting war is frequently observed at around B_2.

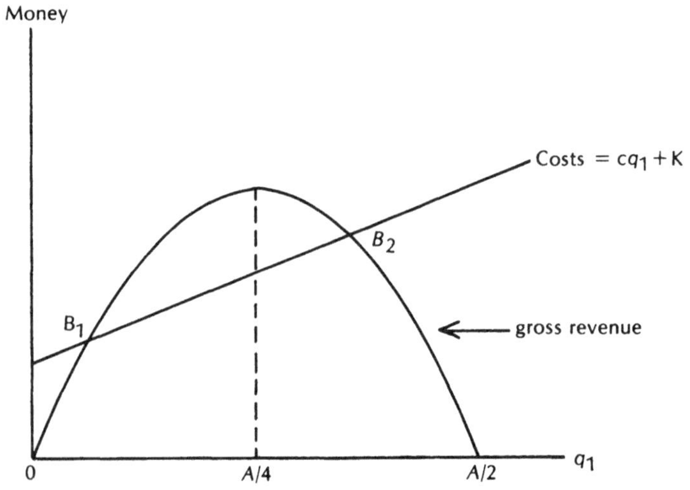

Figure 5.1. Quadratic revenue breakeven chart.

EFFICIENT AND COMPETITIVE SOLUTIONS

If firms were to ignore their influence on the market then each could take price as given. Thus, the *efficient market solution* or price parameter solution is obtained by

$$\max_{q_i} P_i = q_i(p - c), \qquad \text{for} \quad i = 1, 2, \tag{5.10}$$

and

$$p = \frac{\alpha - (q_1 + q_2)}{\beta}.$$

As long as $p > c$, it will pay both firms to produce to capacity. When $p = c$, the firms are indifferent as to their level of production. When $p < c$, both firms would produce nothing.

The symmetric solution for $p = c$ is given by

$$q_1 = q_2 = \frac{\alpha - \beta c}{2}, \tag{5.11}$$

and

$$P_1 = P_2 = 0. \tag{5.12}$$

The solution for the efficient price and production is merely the special case of the general rule for firms with nondecreasing average costs: produce until marginal costs equal market price, or capacity is reached, if that is sooner.

If we set $k_1 = k_2 = (\alpha - \beta c)/2$, the firms together are just able to satisfy the market when $p = c$ and each makes a profit of $P_1 = P_2 = 0$. This has the immediate interpretation that the existing firms have enough capacity to satisfy the demand in a competitive market. Unless a new entrant could come in with lower costs there is no incentive to enter. The shadow price[4] for an extra unit of capacity is zero. We can call this case the *competitive market solution* as entry has been accounted for implicitly.

If $k_1 + k_2 < \alpha - \beta c$, price will be greater than c and extra capital will be of value. Suppose $k_1 + k_2 = (\alpha - \beta c) - \Delta$, then

$$p = \frac{\alpha - (\alpha - \beta c) + \Delta}{\beta} = c + \frac{\Delta}{\beta}, \tag{5.13}$$

and the shadow price or marginal value of an extra unit of capacity is

4. In a more general piecewise linear programming treatment of the multiproduct firm, the value of an extra unit of capacity can be determined from solving the dual program which attaches shadow prices to the firm's productive resources. (See for example Gale, 1960.)

Δ/β. If this value were high enough in a market with unconstrained entry we would expect either capacity expansion of the existing firms, the entry of new firms, or both.

Production Leader Solution

Suppose that for some strategic reason, such as a disclosure law or a time lag in planting a crop, one firm always knows the production plans of the other before it has to select its production level. Furthermore, assume that both firms are aware of the lack of symmetry in information, but it cannot be stopped. This situation is as illustrated in Figure 5.2. The first firm[5] P_1 moves, then the second firm P_2, given this information, selects its production.

If P_2 is given q_1, it will select q_2 yielding

$$q_2^* = \max_{q_2} q_2 \frac{[\alpha - (q_1 + q_2)]}{\beta} - cq_2, \tag{5.14}$$

or

$$q_2^* = \frac{(\alpha - c\beta) - q_1}{2}; \tag{5.15}$$

5. Note the symbol P_i is used to indicate either firm i or the payoff to firm i.

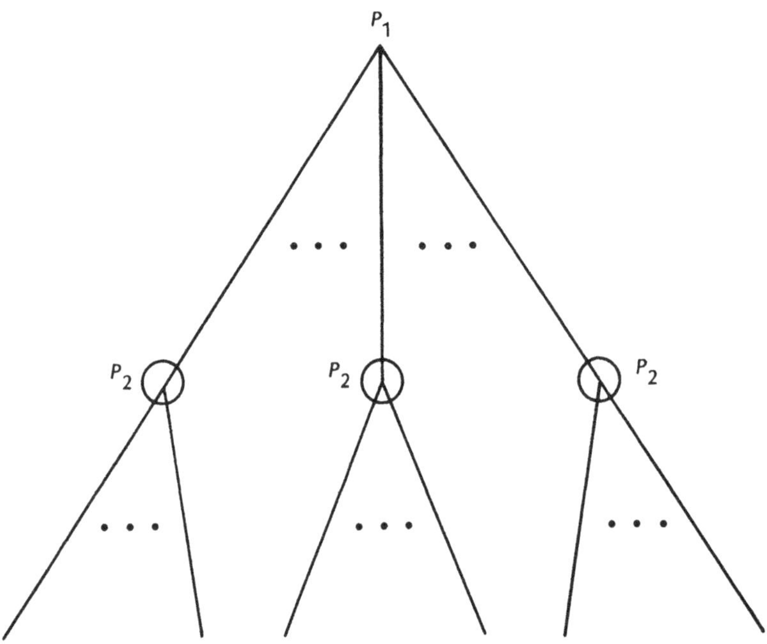

Figure 5.2. Duopoly with perfect information.

hence q_1^* can be calculated from

$$q_1^* = \max_{q_1} q_1\left(\alpha - \left(q_1 + \frac{(\alpha - c\beta) - q_1}{2}\right)\right) - cq_1, \tag{5.16}$$

or

$$q_1^* = \frac{\alpha - c\beta}{2} \quad \text{and} \quad q_2^* = \frac{\alpha - c\beta}{4}. \tag{5.17}$$

We have the paradoxical result that in this situation it pays to have one's production plans known. If the second firm is not in a position to communicate some form of long-run threat to the first, then the first stands to gain by being able to announce a large level of production before the second has selected its production.

This result is a classical paradox of the type encountered in game theory and clearly, noted by Schelling (1960), depends upon the communication or lack of communication between the firms.

Profits under production leadership of firm 1 are

$$P_1 = \frac{1}{8\beta}(\alpha - \beta c)^2 \quad \text{and} \quad P_2 = \frac{1}{16\beta}(\alpha - \beta c)^2. \tag{5.18}$$

Joint Maximum Solution

Suppose that both firms were in a position to collude or cooperate. They could strive to achieve a joint maximum, which can be calculated from

$$\max_{q_1} \max_{q_2} (P_1 + P_2). \tag{5.19}$$

The first-order conditions for the joint maximum are

$$\frac{\partial}{\partial q_i}(P_1 + P_2) = 0, \quad \text{for} \quad i = 1, 2, \tag{5.20}$$

or

$$\left(\frac{\alpha - q}{\beta}\right) - \frac{q}{\beta} - c = 0, \tag{5.21}$$

since $P_1 + P_2 = q[(\alpha - q)/\beta] - cq$. If we assume symmetry of the solution $q_1 = q_2$, we have

$$q_1 = q_2 = \frac{\alpha - \beta c}{4}, \quad p = \frac{\alpha + \beta c}{2\beta} \tag{5.22}$$

and

$$P_1 = P_2 = \frac{1}{8\beta}(\alpha - \beta c)^2. \tag{5.23}$$

Anatomy of a Duopolistic Market

We have solved the duopolistic market for the joint maximum, the noncooperative equilibrium, and the price parameter solutions presented in Chapter 4. We also solved the breakeven and production leadership solutions, in which output is the strategic variable controlled by each firm. These many solutions will play an important role in our analysis of oligopolistic markets. Intuitively, they provide a means of obtaining a rough measure of the degree of competition in a market, and possibly more. To begin to see the relationship among the solutions, several diagrams are presented. Prior to this, however, three further solution concepts are examined: the *Pareto optimal surface,* the *threat curve,* and the *damage-exchange-rate* solution or

$$\max_{q_1} \min_{q_2} (P_1 - P_2).$$

The set of Pareto optimal outcomes consists of those outcomes for which it is not possible for any individual to improve his payoff without diminishing the payoff of at least one other individual. In this example, we count only the firms as individuals and regard the customers as part of a mechanism. When we change this assumption the Pareto set changes. Imagine that both firms select q_1 and q_2 to maximize the profits of the first firm subject to achieving some fixed feasible profit for the second firm. Suppose that profit is Π_2. The goals of the firms can be described as

$$\max_{q_1} \max_{q_2} P_1(q_1, q_2) \tag{5.24}$$

subject to

$$P_2(q_1, q_2) = \Pi_2.$$

In Figure 5.3 the curve $P_2(q_1, q_2) = \Pi_2$ is an *isoprofit curve* for the second firm and is illustrated by R_2CS_2.

Similarly if we consider the profit constraint operative for the first firm, we will achieve the same outcome. Using a Lagrangian multiplier λ the constrained maximization problem expressed by Eq. (5.24) can be written

$$\max_{q_1} \max_{q_2} (P_1(q_1, q_2) - \lambda[P_2(q_1, q_2) - \Pi_2]). \tag{5.25}$$

This yields the first-order conditions for a maximum:

$$\frac{\partial P_1}{\partial q_1} - \lambda \frac{\partial P_2}{\partial q_1} = 0 \quad \text{or} \quad \frac{\partial P_1}{\partial q_1} \Big/ \frac{\partial P_2}{\partial q_1} = \lambda, \tag{5.26}$$

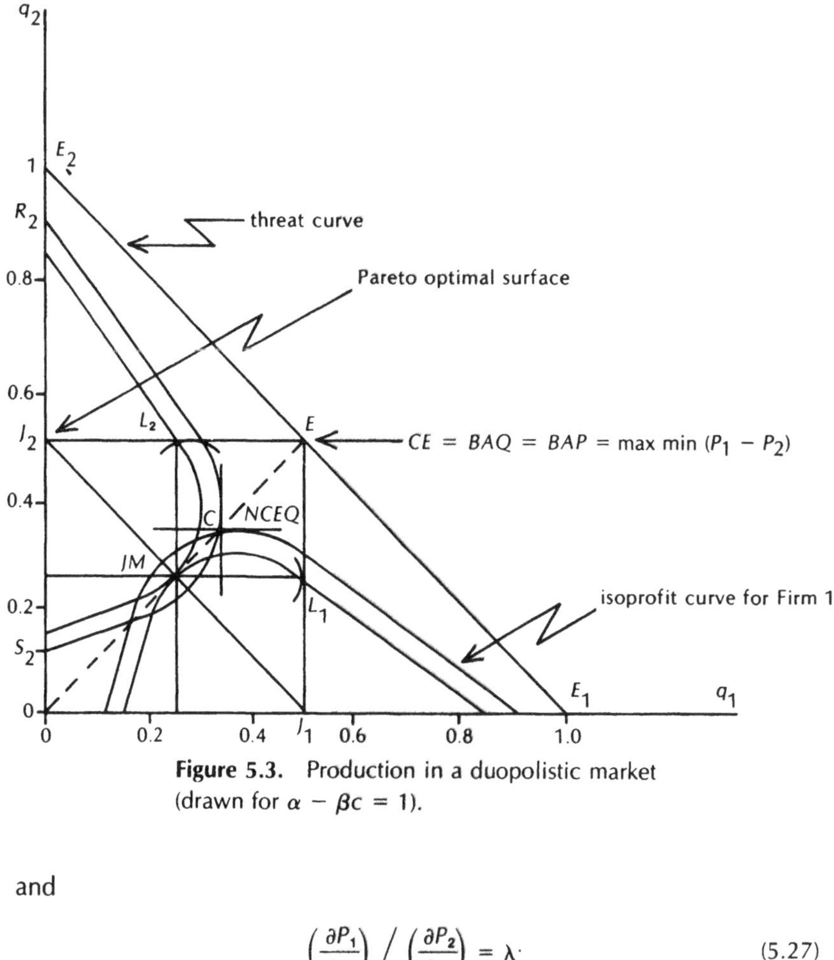

Figure 5.3. Production in a duopolistic market (drawn for $\alpha - \beta c = 1$).

and

$$\left(\frac{\partial P_1}{\partial q_2}\right) \Big/ \left(\frac{\partial P_2}{\partial q_2}\right) = \lambda. \tag{5.27}$$

Equations (5.26) and (5.27) can be expressed as the Jacobian determinant:

$$\begin{vmatrix} \dfrac{\partial P_1}{\partial q_1} & \dfrac{\partial P_1}{\partial q_2} \\ \dfrac{\partial P_2}{\partial q_1} & \dfrac{\partial P_2}{\partial q_2} \end{vmatrix} = 0. \tag{5.28}$$

Calculating the value of this for the market model we obtain

$$[(\alpha - \beta c) - (q_1 + q_2)][(\alpha - \beta c) - 2(q_1 + q_2)] = 0. \tag{5.29}$$

This has two factors each of which is the equation of a straight line. The line

$$q_1 + q_2 = \frac{\alpha - \beta c}{2}, \tag{5.30}$$

which is indicated by J_1J_2 in Figure 5.3, is the Pareto optimal surface. The other line, indicated in Figure 5.3 by E_1E_2, calls for twice as much production and is the threat curve that is discussed below. We note that the point $q_1 = q_2 = (\alpha - \beta c)/4$ is on the Pareto optimal surface and is denoted by *JM*.

In general, for the firms to achieve Pareto optimality they should keep production low. This is not so when the welfare of the customers and the firms are both involved as is shown in Figure 5.6.

Suppose that instead of cooperating, the firms are extremely competitive. Each cares only about besting the other. This amounts to each trying to maximize the difference in profits between them. Solving the first-order conditions for

$$\max_{q_1} \min_{q_2} (P_1 - P_2),$$

we obtain

$$\left(\frac{\alpha - q}{\beta}\right) - \left(\frac{q_1 - q_2}{\beta}\right) - c = 0, \tag{5.31}$$

or

$$q_1 = q_2 = \frac{\alpha - \beta c}{2}, \tag{5.32}$$

which is the efficient point or competitive solution.

When the market is not symmetric the maxmin-the-difference solution does not coincide with the efficient solution. When it is symmetric it always does (see Shubik, 1959b). Thus in this situation extreme competition and a passive attitude ignoring one's own influence on the market lead to the same behavior! In Figure 5.3, the point *E* is marked with both BAQ and BAP. This indicates that the beat-the-average outcome is the same for quantity or price as the strategic variable, as is shown in Figure 5.3.

The *threat curve* or the *Pareto minimal surface* can be calculated in a manner similar to that of the Pareto optimal surface. Given that the first firm tries to maximize its payoff, the difference is that the second now moves to *minimize* the payoff to the first subject to the condition that $P_2(q_1, q_2) = \Pi_2$. This is expressed as

$$\max_{q_1} \min_{q_2} P_1(q_1, q_2), \tag{5.33}$$

subject to

$$P_2(q_1, q_2) = \Pi_2.$$

This leads to precisely the same first-order conditions found in the Pareto optimal surface. The Jacobian determinant given in Eq. (5.28) is obtained. The difference lies in the second-order conditions needed to establish a maximum or minimum.[6] The threat curve is shown by E_1E_2 in Figure 5.3.

The efficient point lies on the threat curve at E, which is the only attainable point on the threat curve if capacity is limited to $k_1 = k_2 = (\alpha - \beta c)/2$. If more capacity is available, one firm's decision to produce q_1 causes the other firm to produce q_2 such that $q_1 + q_2 = \alpha - \beta c$, thereby forcing a profit of zero to each. If excess capacity is so large that one firm can completely flood the market to force price below cost, then a more dynamic analysis of threat is needed.

The noncooperative equilibrium, denoted by C in Figure 5.3, lies at the intersection of the two isoprofit parabolas and is represented by

$$q_i[(\alpha - \beta c) - (q_1 + q_2)] = \frac{(\alpha - \beta c)^2}{3} \quad \text{for} \quad i = 1, 2. \qquad (5.34)$$

The geometrical meaning of the noncooperative equilibrium can be seen from these isoprofit curves. Given the production of one firm as fixed, no action by the other can achieve a higher profit for that firm at the point C.

The Pareto optimal surface and the threat curve are the *loci* of the two families of points of tangency between two sets of isoprofit parabolas. If we chose a nonsymmetric example or one without linear average costs and demand these two curves would not necessarily be straight lines (Shubik, 1959b).

The points L_1 and L_2 indicate productions when firms 1 and 2 are production leaders.

The Pareto optimal surface will no longer be flat even in the example illustrated in Figure 5.3 if $k_i < (\alpha - \beta c)/2$; in this case it is no longer possible for a single firm to supply the optimum monopolistic demand and the points J_1 and J_2 will move toward the origin although JM will not move as long as $k_1 + k_2 \geq (\alpha - \beta c)/2$.

Figure 5.3 displays information concerning profits. The parabolas here represent isoquants, that is, the profit obtainable when $q_1 + q_2 =$ constant. The whole threat curve E_1EE_2 in Figure 5.3 maps into the single point E in Figure 5.4; however, the Pareto optimal surface J_1J_2 maps into J_1J_2 in Figure 5.4.

6. For the Pareto optimal surface the quadratic form derived from the second-order conditions is negative definite, while it is positive definite for the threat curve.

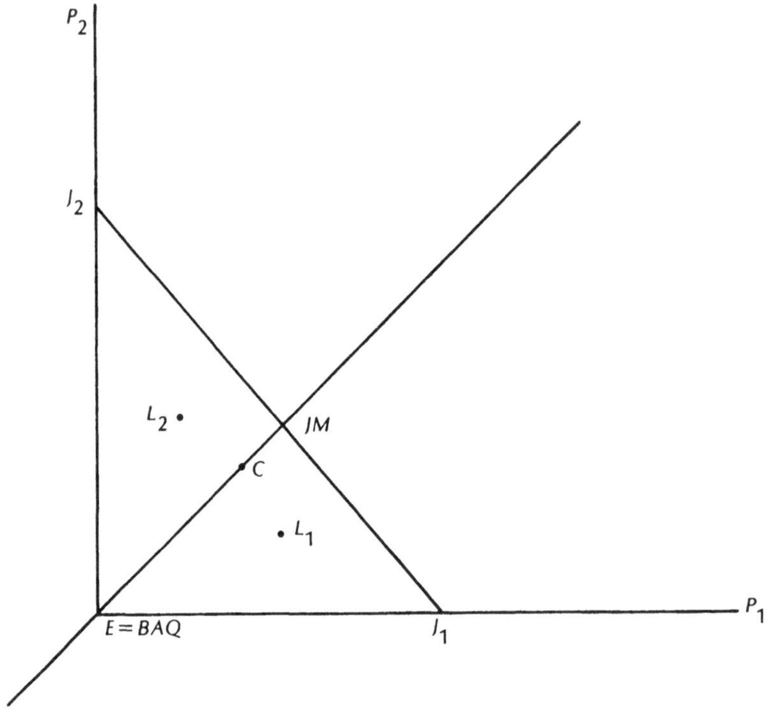

Figure 5.4. Profits in a duopolistic market.

The ratios of levels of production at joint maximum (*JM*), the noncooperative equilibrium, and competitive production are 1 : 4/3 : 2 as is shown in Figure 5.3 and the ratios of profits are 1 : 8/9 : 0 as is shown in Figure 5.4. The points L_1 and L_2 show payoffs from production leadership.

If we wish to consider the consumer's welfare along with the profits of the firms it is necessary to replace the demand function by a utility function. A simple utility function that gives rise to the linear demand is

$$U = \frac{\alpha}{\beta} q - \frac{1}{2\beta} q^2 - pq. \tag{5.35}$$

This function has no income effect (Hicks, 1939), that is, the worth of money does not change to the consumer as his income changes. But for many purchases it is a reasonable approximation to treat wealth in this manner. If this is done, then again only as a first approximation, as Marshall noted, we can calculate a *consumer surplus* as a measure of gain by the consumer in monetary terms resulting from the purchase. Figure 5.5 shows the surplus arising from the purchase of $\bar{q}$ units at price $\bar{p}$. The surplus is given by the area of the triangle *ABC*.

Marshall's argument is that if the firm were able to discriminate against

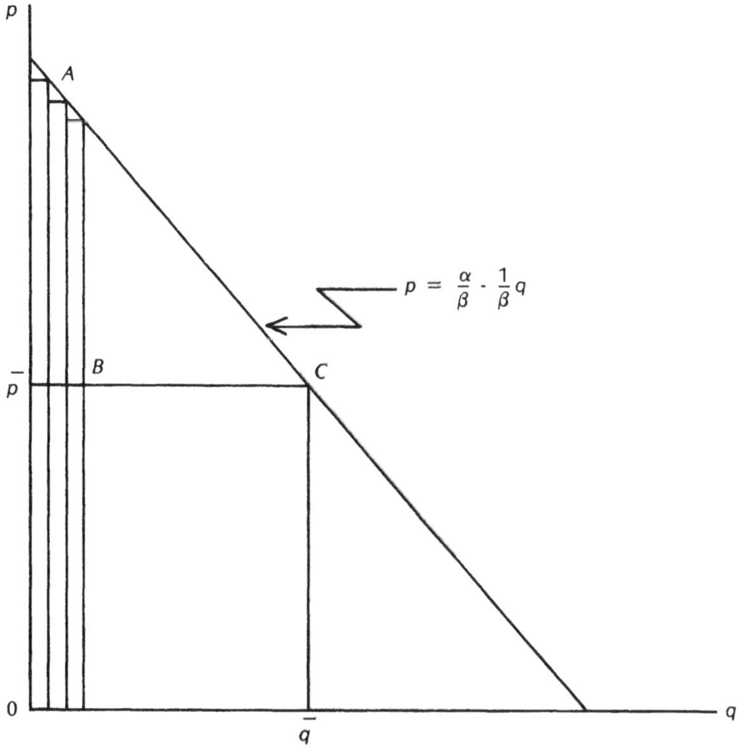

Figure 5.5. Consumer surplus.

the customer it would begin by offering a single unit at a price slightly less than $p = \alpha/\beta$ (see Figure 5.5). The customer would be motivated to buy with virtually no discernible gain in his welfare. The firm now offers an extra increment so that the customer just buys this which he will do if the price is lowered again slightly. This process of perfect discrimination is continued until the point C is reached. By this method of sale the consumer has bought $\bar{q}$ units with the price going from α/β to $\bar{p}$. He has gained virtually nothing (i.e., he has stayed on the same indifference curve even though his assets have changed). If instead the firm offered to sell $\bar{q}$ units at the price $\bar{p}$ then the gain to the consumer would be

$$\int_{q=0}^{\bar{q}} \left(\frac{\alpha}{\beta} - \frac{1}{\beta}q\right) dq - \bar{p}\bar{q} = \left(\frac{\alpha}{\beta} - \bar{p}\right)\frac{\bar{q}}{2}. \tag{5.36}$$

Figure 5.6 illustrates why E' is the efficient point. Although it does *not* lie on the Pareto optimal surface of the firms alone (i.e., the line J_1J_2 in Figures 5.3, 5.4, or 5.6) it does lie on the three-dimensional Pareto optimal surface that indicates the welfare of the consumer and the firms.

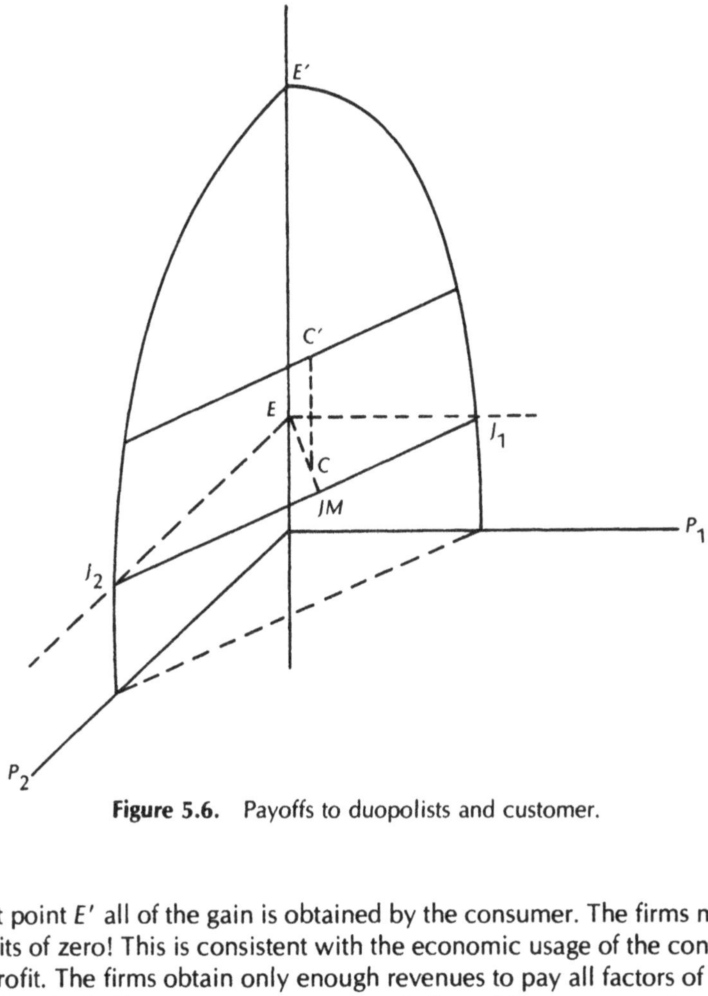

Figure 5.6. Payoffs to duopolists and customer.

At point E' all of the gain is obtained by the consumer. The firms make profits of zero! This is consistent with the economic usage of the concept of profit. The firms obtain only enough revenues to pay all factors of production (including management and capital); nothing more remains, that is, there are no monopolistic gains.

At joint maximum for the firms the payoffs are $[(1/8\beta)(\alpha - \beta c)^2, (1/8\beta)(\alpha - \beta c)^2, (1/8\beta)(\alpha - \beta c)^2]$. At the Cournot equilibrium, they are $[(1/9\beta)(\alpha - \beta c)^2, (1/9\beta)(\alpha - \beta c)^2, (2/9\beta)(\alpha - \beta c)^2]$, where the point C' projects down to C in the payoffs of the firm only. At the efficient point, payoffs are $[0, 0, (1/2\beta)(\alpha - \beta c)^2]$. Point E' projects to E, which shows zero profits for the firms.

In Figure 5.6 it may be seen that C' lies on the Pareto optimal surface for the economy consisting of the firms and the consumer. This is not always true; if the firms had different costs it would not be so.

Duopoly with Identical Products: Price Strategies

Bertrand's Solution

Suppose that the same two firms in the previous section competed by way of price rather than production. Assume that each is endowed with a capacity of at least $k_i \geq \alpha - \beta c$. Each firm has enough product to flood the market, that is, to supply the whole market at a price of $p = c$ or lower. One can see that in this case the noncooperative equilibrium, the efficient point, and maxmin-the-difference solutions all coincide.

With identical products, a customer will prefer to buy from the cheaper firm if there is a price difference. If prices are identical the consumer is indifferent toward the two firms. In Figure 5.7 the line *DF* represents the total market. The line *DD′* is the market share of one firm if they both charge the same price.

Suppose that the first firm charges a fixed price of p_1. How will the demand to the second firm vary as it changes its price from $p_2 = \alpha/\beta$ to $p_2 = 0$? The broken curve *Ddsd′F* shows the relevant contingent demand curve (Shubik, 1959b). When $p_2 > p_1$ the first firm captures the whole market and has enough capacity to satisfy all the demand. Hence the first part of the contingent demand is *Dd*. When $p_2 = p_1$ we assume that the firms split the market at the point *s*. When $p_2 < p_1$ the second firm inherits the whole market. Hence the contingent demand is *d′F*.

The only equilibrium is when $p_1 = p_2 = c$. At this point no one is making a profit or loss. Neither firm is motivated to cut prices as it would make a loss by doing so. Yet neither firm is motivated to raise prices as it would lose all of its market to the other.

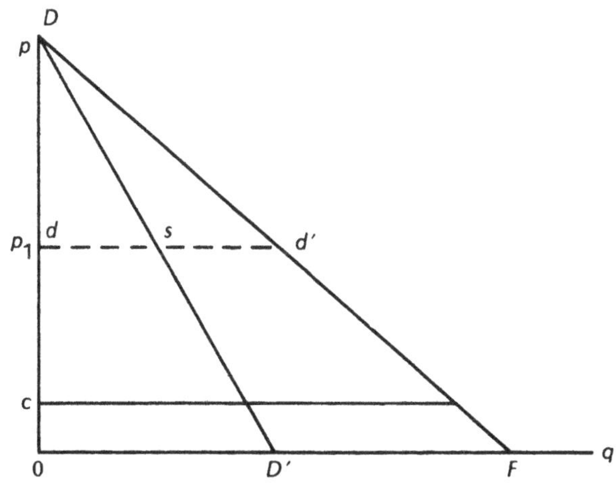

Figure 5.7. Contingent demand in price duopoly.

At any higher price than $p_i = c$ it always pays one firm to slightly undercut the other. Hence no price above c can be in equilibrium. At any price below c the firms are losing money on each sale, and are motivated to raise price to at least c.

EDGEWORTH'S SOLUTION

Suppose the firms each have limited capacity such that neither is able to supply the total market when price equals cost, or $k_i < \alpha - \beta c$. Edgeworth argued that there would be no equilibrium under this circumstance, rather that price would fluctuate over a range (Edgeworth, 1925). If each firm has capacity equal to half of the market at the price $p_1 = p_2 = c$ or if $k_1 = k_2 = (\alpha - \beta c)/2$, the price fluctuation will be between the competitive equilibrium and the individual monopoly prices. This is illustrated in Figure 5.8.

Suppose that both firms charge $p_1 = p_2 = c$. At that point they have just enough capacity between them to supply the market with the amount of product indicated by ce. If the first firm raises its price to p_1^* while the remaining firm is charging $p_2 = c$, the demand curve faced by the remaining firm is given by $\bar{p}_1 d'$; this is in contrast to the zero demand in the Bertrand model. The demand is given by $\bar{p}_1 d'$ because at $p_2 = c$ the capacity of the second firm will be exhausted after supplying $d'e$, leaving

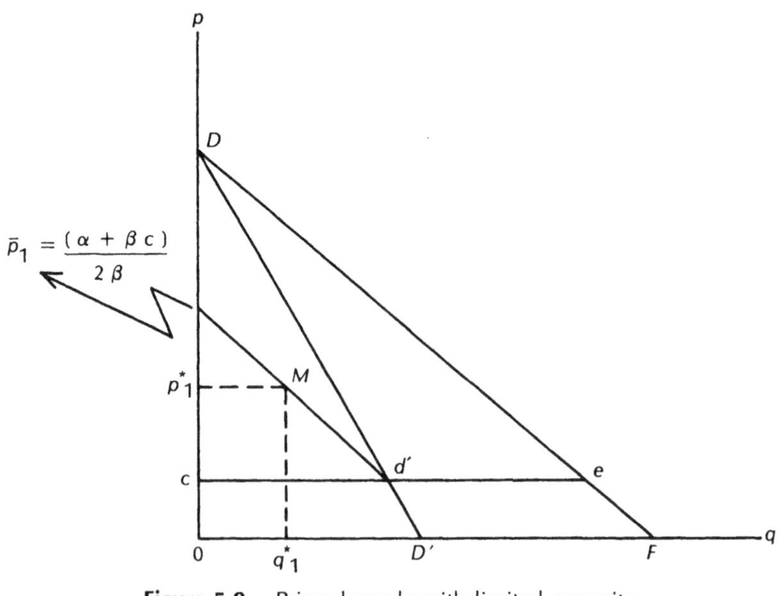

Figure 5.8. Price duopoly with limited capacity.

the first firm with a virtual monopoly in the remaining segment of the market.[7]

The first firm can maximize its profit by raising its price to p_1^* which is the monopoly price against the demand curve $\bar{p}_1 d'$. It will make a profit of $(p^* - c)q^*$. The argument from this point forward must be quasidynamic. Although time periods are unspecified at this point and discounted income streams are not calculated, we contemplate that during the next time period, that is, after the second firm has had a chance to produce again, it will raise its price to just under p_1^*. At that price it will have enough capacity to supply almost all of the demand leaving the first firm with virtually no market unless it cuts its price.

In one form or the other price cutting will continue until a point with $p > c$ is reached such that one of the firms may be motivated to raise price again to p^*. In many markets it is doubtful if the customers would be happy to stand by while price is treated as a yoyo. However, Edgeworth's purpose was to show that under sufficiently increasing average costs of production (or equivalently an appropriate capacity constraint) an equilibrium would not exist. If one is unhappy with the Edgeworth result, the answer is to build a better and richer model of the market, and, to some extent, this is what Chamberlin did.

Chamberlin's contribution to this analysis was to consider differentiated products and entry. In Chapter 6 we examine how these solutions are affected by the first of these additional features.

In summary, the Bertrand and Edgeworth models are consistent. They differ in the treatment of capacity (or increasing average costs). The Bertrand model is a noncooperative game and it happens that the noncooperative equilibrium point coincides with the efficient point when the goods are perfect substitutes. The Edgeworth study indicates that unless there is considerable overcapacity (in this case each firm must be able to supply the whole market at $p = c$), no equilibrium may exist.[8] In Chapter 8 we return to a more detailed consideration of capacity limitations.

Price Leader Solution

Suppose that one firm were required to announce its price first knowing that it could be slightly undercut by its competitor. We see

7. What remains of the market depends on the order in which customers are served. This is discussed in Chapter 11.

8. This model may be reinterpreted in terms of general equilibrium with two sellers with endowments $((\alpha - \beta c/2), 0)$ each, and one buyer with endowment $(0, M)$ where utility functions for the sellers are $cx_1 + y_1$; $cx_2 + y_2$ and for the buyer $\alpha x_3 - (\beta/2)x_3^2 + y_3$; where $y_1 + y_2 + y_3 = M$ and $px_3 = M - y_3$. Thus the instability suggested by Edgeworth is present in a general equilibrium context.

immediately that if the second firm has enough capacity to supply the whole market at $p - c$, the noncooperative price equilibrium coincides with the competitive equilibrium at $p = c$. If the capacity of each of the firms is $k_i > \alpha - \beta c$, the equilibrium must be at $p_1 = p_2 = c$.

Suppose that $k_1 < \alpha - \beta c$ and $k_2 \geq \alpha - \beta c$. If the first firm is the price leader, it names a price $p_1 > c$. The second firm can obtain $(\alpha - \beta p_1)(p_1 - c)$ by undercutting slightly, leaving the first with nothing. The second firm has the option of naming a high price, letting the first firm sell k_1 at price p_1. The second firm can obtain

$$\max_{p_2} (\alpha - \beta p_2 - k_1)(p_2 - c),$$

implying that

$$p_2 = \frac{\alpha + \beta c - k_1}{2\beta} > p_1.$$

It is then in the interest of the price leader to name as high a price as possible, such that the second firm will name a higher price rather than undercut. This will be

$$(\alpha - \beta p_1)(p_1 - c) = \frac{1}{\beta}\left(\frac{\alpha - \beta c - k_1}{2}\right)^2. \tag{5.37}$$

If the first firm has capacity $k_1 \geq \alpha - \beta c$ and the second firm has less, then the first must expect to be undercut. Hence it should select

$$p_1 = \frac{\alpha + \beta c - k_2}{2\beta}. \tag{5.38}$$

While these solutions may be formally correct under static situations, they are highly unrealistic. This emphasizes the need to provide a truly dynamic analysis for a price leader duopoly. A different interpretation of a price leader model is one in which the first firm sets price and the second is required to accept that price as given.

Other Duopoly Models with Identical Products

Price Quantity Model

In the short run a firm may have to commit itself to a level of production and a price. Thus, both price and production may be considered as independent or strategic variables in a world where, at least in the short run, excess inventories and stockouts may be encountered.

When both firms each have capacity greater than or equal to $\alpha - \beta c$ and no costs to carrying inventories, it is easy to establish that the strategies $p_1 = p_2 = c$ with $q_1 = q_2 = (\alpha - \beta c)/2$ form an equilibrium point. If the firms have capacities smaller than $\alpha - \beta c$ or if there are inventory costs, no pure strategy equilibrium may exist.

If no specific set of choices by the firms leads to an equilibrium we may wish to consider a different solution concept. A mathematically interesting extension of the equilibrium point, but one that is somewhat hard to justify in a static economic model, is the mixed strategy equilibrium point. This is discussed further in Chapter 8.

Other Models

It is formally possible to consider markets in which two firms have different strategic variables. Thus we could consider that one firm independently sets price and production, while the other sets only production. In this listing of different solutions to static duopoly with identical products, no attempt has been made to cover the many cooperative solutions, such as the core, value, or nucleolus of the cooperative market. These are partially discussed elsewhere (Shapley and Shubik, 1971–74). It must also be noted that the distinction between markets with sidepayments and markets without sidepayments has not been made.

A market with sidepayments is one in which firms can collude to exploit the market and then make money payments to each other to adjust the distribution of their profits from collusion. When the firms are symmetric, they need not make direct money payments to each other when colluding. They can achieve all profit adjustments by controlling production and market price. The ability to make sidepayments is important for cartels with members of different sizes and with different costs. Inefficient members can be paid to cut back production. If sidepayments are forbidden, this alternative is not available.

Price and production are by no means the only strategic variables of an oligopolistic firm. They are, however, so universally present in economic competition that they provide the natural first set of models to be considered.

6 | Duopoly with Product Differentiation

BEFORE THE ANALYSIS OF DUOPOLY with differentiated products is presented, the structure of oligopolistic demand must be investigated. A general discussion of oligopolistic demand is presented in Appendix B; attention is limited here to the two-person case that can be illustrated geometrically and algebraically.

Contingent Demand

When two individuals sell differentiated products and charge different prices, how much will they sell? This is, in general, a problem in general equilibrium analysis[1] with rationing (unless the inventories of the firms are assumed to be always sufficient for the demand arising from any contingency). Given a fixed price $\bar{p}_2$ for the second firm, the contingent demand for the first firm may be expressed as $q_1 = f(p_1|\bar{p}_2)$. To make this problem amenable to a partial equilibrium analysis as a first approximation, the income effect (see Chapter 5) of different prices for this set of competing commodities is ignored.

We assume that the preferences for the consumers considered in ag-

1. General equilibrium analysis deals with an economy described as a complete feedback system; all repercussions of economic activity are taken into account. Partial equilibrium studies a group of agents embedded in an environment that does not necessarily include all feedbacks. Thus, for example, in the study of oligopoly frequently the customers are an outside mechanism. They are not modeled directly as part of the system.

gregate can be represented by a quadratic function:

$$U = \frac{\alpha}{\beta} q - \frac{1}{2\beta} q^2 - \frac{2\sigma^2}{\beta(1 + \gamma)} - \sum_{i=1}^{2} p_i q_i, \tag{6.1}$$

where

$$\sigma^2 = \left[\frac{q_1 - q_2}{2} \right]^2$$

and γ is a measure of the substitutability of the goods. When supplies are sufficient and prices are not too far apart (this will be made precise below) the demand conditions for both products are derived by solving

$$\frac{\partial U}{\partial q_1} = \frac{\partial U}{\partial q_2} = 0 \tag{6.2}$$

simultaneously, which yields

$$\frac{\alpha}{\beta} - \frac{q}{\beta} - \frac{(q_1 - q_2)}{\beta(1 + \gamma)} = p_1 \tag{6.3}$$

and

$$\frac{\alpha}{\beta} - \frac{q}{\beta} - \frac{(q_2 - q_1)}{\beta(1 + \gamma)} = p_2. \tag{6.4}$$

Adding Eqs. (6.3) and (6.4), we obtain

$$\frac{\alpha}{\beta} - \frac{(q_1 + q_2)}{\beta} = \frac{p_1 + p_2}{2}; \tag{6.5}$$

subtracting these same equations yields

$$-\frac{1}{\beta(1 + \gamma)} (q_1 - q_2) = \frac{p_1 - p_2}{2}. \tag{6.6}$$

From Eqs. (6.5) and (6.6), the demands q_1 and q_2 can be obtained in terms of prices p_1 and p_2. They are

$$q_1 = \frac{1}{2} \left[\alpha - \beta \left(1 + \frac{\gamma}{2} \right) p_1 + \frac{\beta \gamma p_2}{2} \right] \tag{6.7}$$

and

$$q_2 = \frac{1}{2} \left[\alpha + \frac{\beta \gamma p_1}{2} - \beta \left(1 + \frac{\gamma}{2} \right) p_2 \right], \tag{6.8}$$

where p_1, p_2 and q_1, $q_2 \geq 0$.

Figures 6.1 and 6.2 illustrate the key aspects of contingent demand in a duopolistic market for all price and capacity possibilities. These diagrams should be compared with Figures 5.7 and 5.8 which illustrate the case of products as perfect substitutes. We note from comparison that unlike the market with perfect substitutes there is no longer a discontinuity encountered when prices are equal, hence we need no special (and more or less arbitrary) rule to determine the point s.

The following discussion is based on the price changes of the first firm, assuming the price for the second firm is fixed at p_2^* and the second firm always has enough inventory to meet its demand. The axes in Figure 6.1 are labeled for the variables of the first firm. Given $p_2 = p_2^*$, the contingent demand equation for the first player is given by Eq. (6.7) in the range dd'. Beyond d and d' other phenomena are encountered and must be examined.

Equations (6.7) and (6.8) are obtained by solving Eqs. (6.3) and (6.4) simultaneously. However, implicitly we require that q_1 and $q_2 \geq 0$. In other words, we do not wish to consider negative demands! At the point d the first firm has raised its price sufficiently above the price of the second firm to price itself out of the market. This point can be easily calculated from Eq. (6.7) by setting $q_1 = 0$ and $p_2 = p_2^*$, which yields

$$p_1 = \left(\frac{\alpha}{\beta} + \frac{\gamma}{2} p_2^*\right) \bigg/ \left(1 + \frac{\gamma}{2}\right). \tag{6.9}$$

Now suppose that the first firm undercuts the second. At prices slightly less than p_2^* the market for the first firm will increase fast along dd'. The second firm has plenty of customers to lose. However, at the point d' the price has been reduced sufficiently such that the second firm is priced out

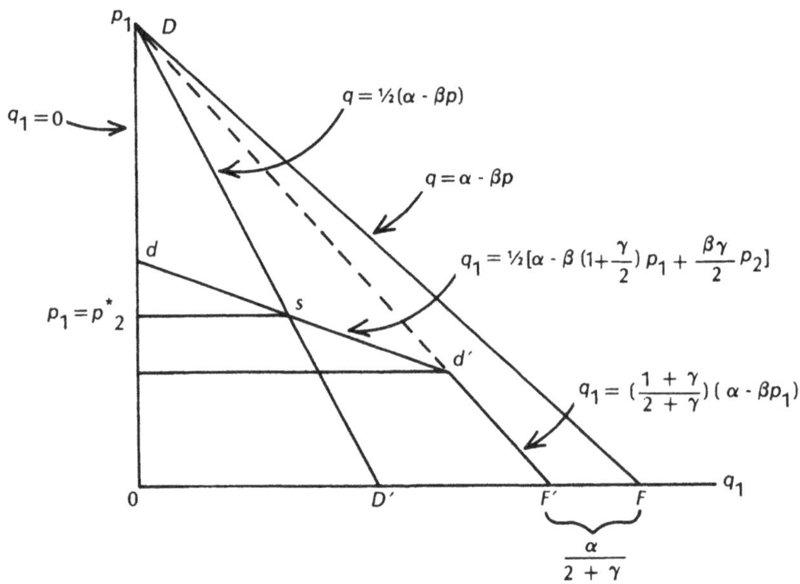

Figure 6.1. Contingent demand with differentiated products.

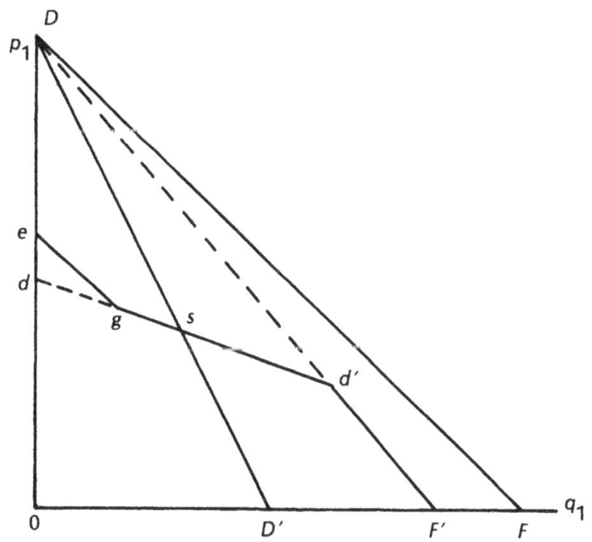

Figure 6.2. Contingent demand with a capacity constraint.

of the market, or $q_2 = 0$. This is easily calculated by setting $q_2 = 0$, $p_2 = p_2^*$ and solving for p_1 in Eq. (6.8), yielding

$$p_1 = \left[\left(1 + \frac{\gamma}{2}\right) p_2^* - \frac{\alpha}{\beta}\right] \Big/ \frac{\gamma}{2}. \tag{6.10}$$

This requires also that $p_1 > 0$; otherwise no satisfactory price exists.

Although the second firm has been driven out of the market at d', the first firm could continue to cut price if it so wished. It will not increase its number of customers by much since the second firm no longer has customers to lose. The second branch $d'\,F'$ of contingent demand can be calculated from Eq. (6.3) by setting $q_2 = 0$. This gives

$$p_1 = \frac{\alpha}{\beta} - \frac{1}{\beta}\left(\frac{2 + \gamma}{1 + \gamma}\right) q_1. \tag{6.11}$$

The distance $F'F$ has yet to be interpreted: it measures the degree of nonsubstitutability between the two goods. By setting $p_1 = 0$ in Eq. (6.11) and observing that at F, $q = \alpha$, the distance $F'F$ is shown to be $\alpha/2 + \gamma$. When $\gamma = 0$ this is $\alpha/2$ which means that F' has moved over to D' and DD' is the total contingent demand curve since the firms have become isolated monopolists. When $\gamma \to \infty$, $F'F \to 0$. As the products approach complete substitutability $d'\,F'$ approaches DF and dsd' becomes horizontal. This situation should be compared with that illustrated in Figure 5.6.

Similar to the case presented above is that of the capacity limitation of the firm. Suppose that the second firm has a capacity limit of k_2. It can be seen from Figure 6.2 that an extra "kink" in the contingent demand may be introduced by this capacity constraint. Figure 6.2 is identical to Figure 6.1 with the exception of the segment *ge* where *g* lies on *sd*. At point *g* the lack of capacity of the second firm affects the demand for the product of the first. If the second firm had more capacity, the first firm would continue to lose sales along *dd'* as it raised its price. However, the second firm can no longer supply the customers who wish to switch. The point *g* can be calculated by setting $q_2 = k_2$ and $p_2 = p_2^*$ in Eq. (6.8). We obtain

$$p_1 = \frac{2}{\beta\gamma}\left[(2k_2 - \alpha) + \beta\left(1 + \frac{\gamma}{2}\right) p_2^*\right]. \tag{6.12}$$

The segment *ge* is calculated directly from Eq. (6.3) setting $q_2 = k_2$. This gives

$$p_1 = \frac{\alpha}{\beta} - \frac{1}{\beta}\left[\frac{(2 + \alpha)q_1 + \gamma k_2}{1 + \gamma}\right]. \tag{6.13}$$

Linear segments are obtained in this example because it is a quadratic utility function, which enables us to carry out specific calculations. Although these calculations are relatively simple in this instance an application to calculate the contingent demand structure for *n* firms selling nonsymmetrically related products becomes quite difficult as shown in Appendix B.

Figure 6.3 shows a more general shape for a contingent demand. The importance of *dsd'* is that it may easily contain kinks and inflections. The importance of these, as is noted in the next section and elsewhere (Shubik, 1959b), is that the existence of market equilibrium may depend upon them.

Leaving aside strategies involving product variation and restricting action to price and production if there are *n* firms, each contingent demand depends in detail on $2n - 1$ variables, that is,[2]

$$q_i = f(p_1, p_2, \ldots, p_n; q_1, \ldots, q_{i-1}, q_{i+1}, \ldots, q_n).$$

Fortunately the situation may often be simplified. When firms have enough capacity and produce in sufficient quantity, beyond some level their production will no longer influence the contingent demand of the others.

The empirical problem in the study of contingent demand conditions is to find a sufficiently accurate aggregation or approximate formula that can be estimated and used in the approximate estimation of demand in the region currently of interest. For example, in a symmetric game, only

2. The program for allocating demand in the oligopoly game actually carries out computations on this basis.

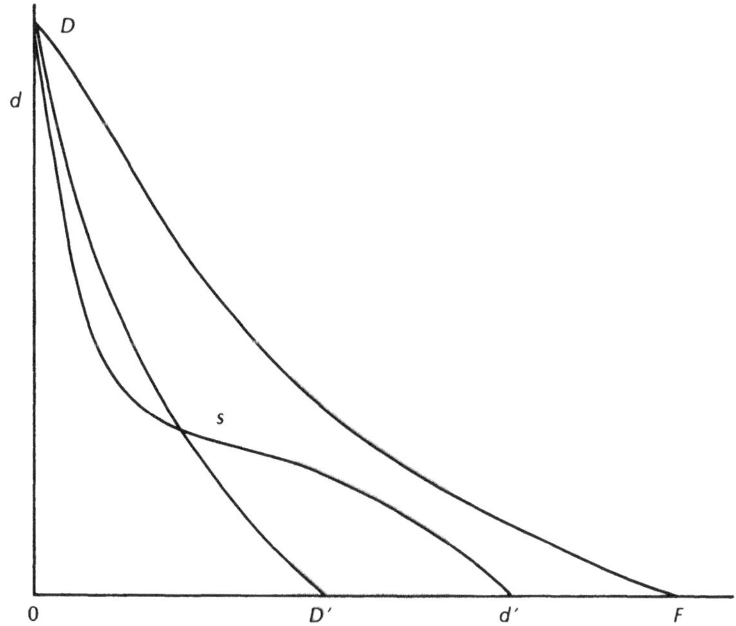

Figure 6.3. A general contingent demand.

the average of all others' prices is needed to estimate demand in the region of equilibrium. This amounts to estimating one of the family of curves denoted by *dd′* in the neighborhood of an appropriate *s*. This is illustrated in Figure 6.3.

In daily operations and even in theorizing, which is not highly mathematical, we do not regard oligopolistic or contingent demand functions as functions of $2n - 1$ variables. We may be correct in not doing so for reasons other than simplicity. When dynamic aspects of markets are included a different and more behavioral model of both consumers and firms may be called for.

Even if we were to acknowledge the effect of all other prices and production levels in the study of market demand, the problem does not stop there. The aggregate market demand function discussed by economists is comprised of many different individual demands, as will be pointed out by any marketing or advertising consultant worth his salt. There are many different patterns of individual demand that give rise to the same aggregate picture. In oligopolistic competition, individual firms actively chase special segments of the market. Hence the aggregate demand view of consumers may not be enough.

In the model constructed here, the simplest of assumptions were used to enable calculation of all contingent demands. Other assumptions could have been made and would have given different contingent demands. For example, in some markets, snob or Veblen effects (Veblen, 1899) might send some customers to the firm with the higher price because the price is regarded as a symbol of the quality. In France, for example, a form of conspicuous consumption has been reached where items carrying the advertising of a firm command a premium over those that do not. Thus, a silk scarf with the name Balenciaga or Dior worked into the design in a conspicuous manner is valued by some customers far more than the scarf with the advertising omitted.

The study of the profile of customers and the details of how they are served is an important part of marketing and advertising. It is also an important aspect of economic theory and is of direct importance in obtaining an accurate model of market demand in any oligopolistic market.

Symmetric Duopoly with Product Differentiation

The model presented here and several variants of it designed to account for more than two firms of different size form the basis for the analysis of oligopoly that follows. We are now in a position to examine the analogues of the Bertrand, Edgeworth, and Cournot treatments of duopoly presented earlier for the case in which the products are differentiated.

Price and Price-Quantity Equilibrium

The picture of the market structure in the price-quantity strategy model is of two large firms supplying a mass market via a diffuse and passive retailing system. Since the retailers are regarded as mere mechanisms carrying out the orders of the manufacturers they can be abstracted out of the model. If we assume that each firm sets a price for its particular product we need to state how the market is assigned and what are the supplies available. If we wish to concentrate first on the implications of price strategies, we can assume that either (1) the market is for a "made-to-order" item, that is, first the business is booked, then production takes place; or (2) the firm sets both price and production, but inventory carrying costs and stockout penalties are zero.

Ignoring the conditions of inventory carrying costs and of stockout penalties and assuming that both firms have adequate capacity, we obtain the conditions for the existence of the noncooperative equilibrium:

$$\frac{\partial}{\partial p_i}\left[\frac{1}{2}(p_i - c)\left(\alpha - \beta\left(1 + \frac{\gamma}{2}\right)p_i + \frac{\beta\gamma}{2}p_j\right)\right] = 0, \quad \text{for} \quad i = 1, 2;\ j = 2, 1. \qquad (6.14)$$

Solving this for p_i, we obtain

$$p_i = \frac{\alpha/\beta + c(1 + \gamma/2)}{2 + \gamma/2} \quad \text{for} \quad i = 1, 2, \qquad (6.15)$$

yielding a profit of

$$P_i = \left(\frac{1 + \gamma/2}{2\beta}\right)\left(\frac{\alpha - \beta c}{2 + \gamma/2}\right)^2. \tag{6.16}$$

Leaving aside problems of entry, this noncooperative equilibrium point is basically the same as Chamberlin's (1933, 1962) large group equilibrium. In Figure 6.4 this equilibrium point is described geometrically as the point of tangency between the isoprofit curve *rs* and the contingent demand segment *dd'* at its point of intersection, *s*, with the curve *DD'*. The distance from *s* to the line *cc'* which represents the constant average cost of production gives a measure of the oligopolistic price increase above the competitive price. The rectangle *cfsc'* indicates the oligopoly profit to each firm.

On examining Eq. (6.15) it can be seen that when $\gamma = 0$ the monopoly price as given in Chapter 5 is obtained; when $\gamma \to \infty$, the price under pure competition results.

Capacity Constraints and the Edgeworth Cycle

If both firms do not have sufficient capacity (as γ varies, sufficient capacity will amount to twice that needed to supply the market at efficient production), the equilibrium may not exist and the market will be subject to the instability indicated by Edgeworth. As shown in the previous section especially Figure 6.2, capacity limitation introduces a new "kink" into *dd'*. This is shown in Figure 6.4 as ge. If ge is sufficiently steep to touch or cut the isoprofit curve *rs*, as is indicated by the point *h*, the

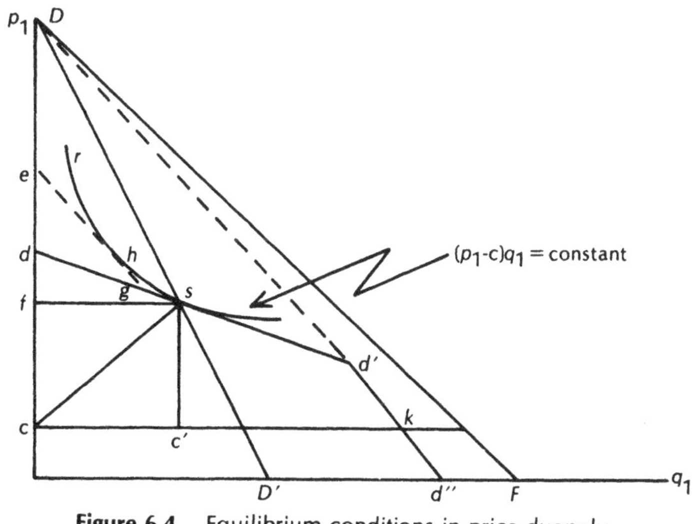

Figure 6.4. Equilibrium conditions in price duopoly.

equilibrium is destroyed and the firms may be motivated to raise price, fluctuating in the range from s to h.

Calculations of capacity limitations and further investigations of the Edgeworth cycle and its relationship to the mixed strategy interpreted as a probability distribution over price are given in Chapter 8. Here we give only the barest outline and indicate the meaning of the range of fluctuations in Figure 6.5 which shows the payoffs to firm 1 as a function of its price, given the price of firm 2.

We can calculate the point at which the new demand curve ge in Figure 6.4 provides enough potential profit to destroy the noncooperative equilibrium at s. Suppose that each firm has more than enough capacity K to provide for its market share at the price noncooperative equilibrium s. If the first firm raises its price it moves along the demand curve dd' until the second firm runs out of capacity. At this point the first firm faces the demand curve ge. When ge touches the isoprofit curve the pure strategy noncooperative equilibrium is destroyed and we can write down the mathematical conditions to solve for K. We equate the profits of the first firm at the noncooperative equilibrium (given in Eq. (6.16)) with its profits as a monopolist facing ge as its demand curve. This yields

$$K = (\alpha - \beta c)\left(\frac{1+\gamma}{\gamma}\right)\left[1 + \frac{1}{2}\left(\frac{2+\gamma}{4+\gamma}\right)\sqrt{\frac{2}{\beta(1+\gamma)}}\right]. \tag{6.17}$$

A quick check shows that this equation is unbounded for $\gamma = 0$ (each firm

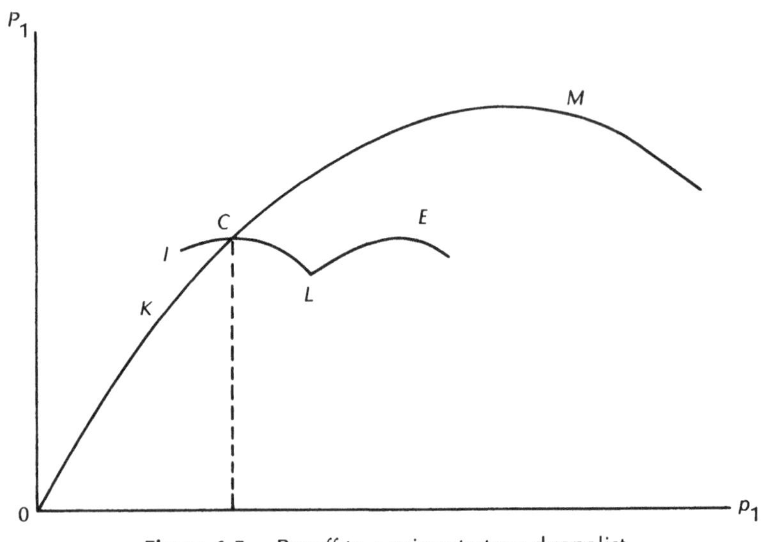

Figure 6.5. Payoff to a price strategy duopolist.

is a monopolist) and for $\gamma \rightarrow \infty$ we obtain $K = \alpha - \beta c$, which implies that each firm must have enough product for the whole market. (See Shapley and Shubik, 1969b, for more detail.)

In Figure 6.5 the curve *OKCM* shows profits if both firms charge the same price (they each face the demand *DD'*). The first section *OK* is linear; it is where they both sell to capacity. Then section *KCM* is parabolic. The curve *ICL* shows the profits of firm 1 as it varies its price, given that firm 2 has its price fixed at the noncooperative equilibrium point *C* (firm 1 faces the demand curve *dd'* in Figure 6.4). At point *L* the form of the demand curve switches from *dd'* to *ge* and, as the first firm continues to raise price, its profits rise again along a new parabola until they reach a maximum at *E* which has the same height as *C*. The Edgeworth cycle is between *C* and *E*.

The Price Leader Solution

It follows from the analysis of the Edgeworth cycle that the bottom of the range gives the price which the firm should select that has to announce price first. In general, this is not the same as the most cautious, or the maxmin policy. The first firm does not assume that the other is out to damage it, but that the other is only concerned with its own profits (see Chapter 6).

Once again we have taken a constrained view of price leadership equating it with naming price first and announcing the price to all. In a vaguer sense, the price leader is meant to be a *leader* in a dynamic process. It is difficult to give this a clear operational mathematical meaning.

In summary, product differentiation provides enough monopoly power for the existence of monopolistic profits. These vary with the degree of differentiation. This alone is not enough to guarantee market stability. Excess capacity is needed. The amount of excess capacity depends upon details of market structure, specifically the shape of the cost functions, number of firms, and the shape of the contingent demand functions.

The Quantity or Extended Cournot Market

Given product differentiation, it is still possible to define a quantity-strategy noncooperative equilibrium. Equations (6.3) and (6.4) give the prices that clear the markets when quantities q_1 and q_2 are named. Hence we can write the noncooperative equilibrium conditions as

$$\frac{\partial P_i}{\partial q_i} = \frac{\partial}{\partial q_i}\left\{q_i\left(\frac{\alpha}{\beta} - \frac{q}{\beta} - \frac{(q_i - q_j)}{\beta(1+\gamma)} - c\right)\right\} = 0,$$

$$\text{for } i = 1, 2;\ j = 2, 1. \qquad (6.18)$$

which yield

$$q_i = (\alpha - \beta c)\left\{\frac{1 + \gamma}{4 + 3\gamma}\right\}. \tag{6.19}$$

We note for $\gamma = 0$ this becomes the monopolistic production level and as $\gamma \to \infty$ it becomes the noncooperative production without product differentiation, or the Cournot output.

Comparing the two noncooperative equilibria we have the price game

$$p_i = \frac{\alpha/\beta + c(1 + \gamma/2)}{2 + \gamma/2}$$

and the quantity game

$$p_i = \frac{\alpha/\beta(1 + \gamma/2) + c(1 + \gamma)}{2 + 3\gamma/2}$$

When $\gamma = 0$, these equations are equal at the monopoly price; however, for $\gamma > 0$ the difference between the two may be expressed as $[(\alpha/\beta) - c](\gamma^2/4)$ which is always positive for $c < \alpha/\beta$. (This is the price at which demand drops to zero; hence costs will be less if the market is to survive.) From this expression we can deduce that the quantity noncooperative equilibrium price in this market is always greater than the price equilibrium when $\gamma > 0$.

Threats with Price Strategies

Even in a static model, it is important to note the effect of capacity on Pareto optimality and threat strategies. The differences in strategic structure imposed by using price instead of production as the strategic variable should be appreciated. In the price-strategy duopoly, the firms cannot achieve many of the outcomes on their Pareto optimal surface that they were able to achieve using production as the strategy. This can be seen immediately when it is observed that with symmetry and perfect substitutes the Cournot duopolists can achieve any payoff they want by adjusting production; however, competitors using price as a strategy (when there is plenty of capacity around) can only achieve the individual monopoly profits (J_1 or J_2 in Figure 6.6) or the joint maximum, or any even split on the line 0 to JM. Figure 6.6 should be contrasted with Figure 5.2 where the quantity-strategy duopolists can achieve any point on J_1J_2.

Depending upon capacity conditions and the degree of substitution between the products, the achievable profits of the two firms may look like the surface $M_1K_1JMK_2M_2$. Even without further investigation we may see the meaning of this unusual diagram in terms of the difference in side-payment possibilities in the price and quantity games. When products are identical the points J_1 and M_1, J_2 and M_2 coincide. When they are dif-

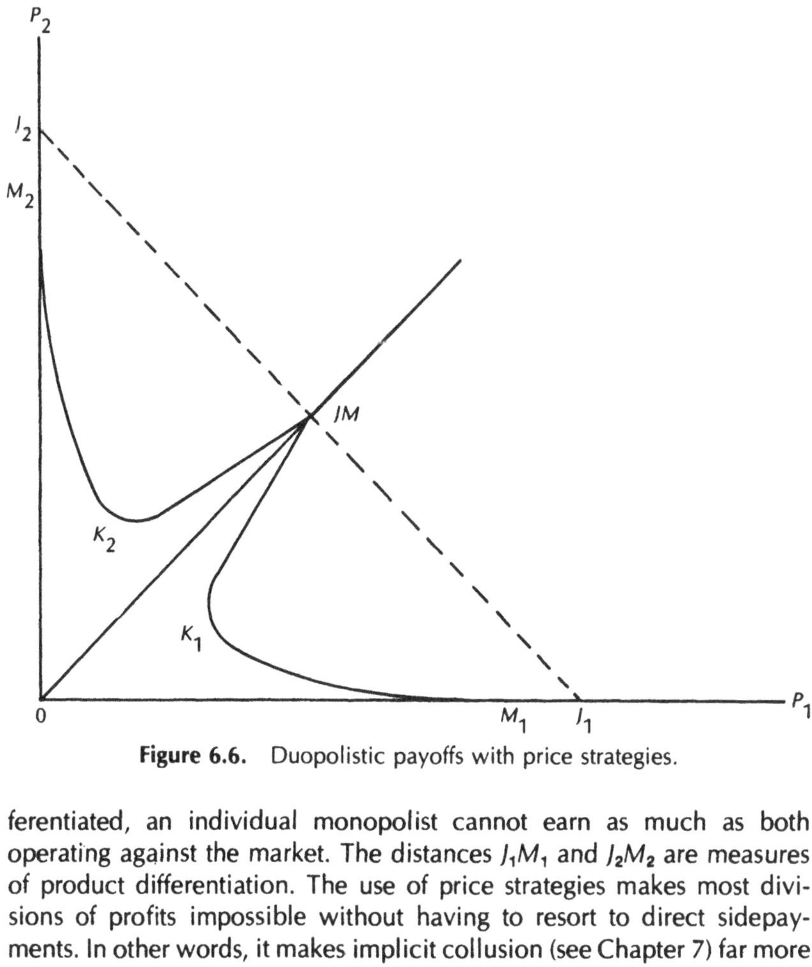

Figure 6.6. Duopolistic payoffs with price strategies.

ferentiated, an individual monopolist cannot earn as much as both operating against the market. The distances J_1M_1 and J_2M_2 are measures of product differentiation. The use of price strategies makes most divisions of profits impossible without having to resort to direct sidepayments. In other words, it makes implicit collusion (see Chapter 7) far more difficult than with production strategies.[3]

Although the solution to maxmin $(P_1 - P_2)$ is the same for both price and quantity strategies in the example of this chapter, this is not always true and the threat possibilities are quite different.

3. The presence of sidepayments means that there is some good or substance in the game such that all individuals have a constant marginal utility for it and that it can be freely transferred among them. A good first approximation in oligopolistic competition is money. When money transfers are forbidden other devices such as jointly coordinating production or exchanging customers may be used. The relationship between collusion and the presence of sidepayments is subtle. One can collude without resorting to overt transfers of wealth among colluders. However if it is legal to make transfers, the feasible set of achievable outcomes is generally enlarged and it becomes easier to "cut the cake" into any size of portions. Explicit laws against money sidepayments do not rule out collusion, they merely may make it more cumbersome.

Care must be taken when using special examples to ensure that they do not obscure important general phenomena. In this instance, the possibility of price discrimination at joint maximum profits has been hidden by the choice of this example. For instance, if it were possible to divide the customers so that a rich or a "snob" segment of the market went to the firm charging the higher price and the rest of the market went to the other firm, it might easily be jointly maximal for the firms to maintain different prices even though their products are symmetric substitutes. This is an example of a nonsymmetric solution for behavior in a symmetric market.

Other Solution Concepts

Several other solution concepts have been suggested. For example, both firms might be paranoid about the intentions of the other. Each might believe that the other wishes to do it as much harm as possible. This can be expressed as

$$\begin{matrix} \max\limits_{q_1} \min\limits_{q_2} P_1 & & \max\limits_{p_1} \min\limits_{p_2} P_1 \\ & \text{or} & \\ \max\limits_{q_2} \min\limits_{q_1} P_2 & & \max\limits_{p_2} \min\limits_{p_1} P_2 \end{matrix} \tag{6.20}$$

depending upon whether there is quantity or price competition.

One firm might believe that the other is out to do it damage and be right. This can be described as

$$\max_{q_1} \min_{q_2} P_1, \quad \text{and} \quad \min_{q_2} P_1 \quad \text{subject to} \quad P_2 \geq 0. \tag{6.21}$$

We have described this solution with production as the strategic variable. In all cases we can specify the same solution concept with one of many different strategic variables.

With only two players it is easy to see that

$$\text{maxmin}\,(P_1 - P_2) \tag{6.22}$$

and beat-the-average or

$$\max \left(P_i - \frac{1}{n-1} \sum_{j \neq i}^{n} P_j\right) \tag{6.23}$$

are equivalent. Finally the firms might compete to maximize market share, but this concept must be qualified before it can be completely well defined, as is noted below.

It is of interest that all of these extra "economic warfare" solutions depend critically upon the capacities of the players. Capacity limits are fundamentally a dynamic phenomenon. They imply the presence of a configuration of capital goods and organization that cannot be changed in the short run. An adequate theory of oligopolistic competition must be dynamic. However, at least an indication of the interlinkage between these static solutions and dynamics is given by introducing capacity constraints.

Beat-the-Average and Related Solutions

For the symmetric game with two firms and sufficient excess capacity it has been shown that the competitive equilibrium or efficient point and the maxmin $(P_1 - P_2)$ solutions coincide. This is true for either price or quantity as the strategic variable. In the symmetric two-person market the beat-the-average solution also coincides with the max min $(P_1 - P_2)$ solution. This means that six conceptually different models predict the same outcome in this symmetric market! Three solution concepts each solved with either price or quantity as the strategic variable all lead to the same result. This result shows the strength of the outcome as a predictor (many theories predict a single outcome) and at the same time show the danger of studying symmetric models unless one is extremely careful. For some markets five of these models would predict different outcomes.

When there is insufficient excess capacity there is no stable equilibrium to the $\max_{p_1} \min_{p_2} (P_1 - P_2)$. This is shown elsewhere (Shubik, 1959b).

Variants of Maxmin Solutions

The two solution concepts noted in Eqs. (6.20) and (6.21) do not give the same results in the price and quantity strategy variants. The price game is calculated first. In all instances we assume that the firms have at least sufficient capacity to satisfy the market at $p_1 = p_2 = c$. We solve for capacities in the range $(\alpha - \beta c)/2 \leq k_1, k_2 \leq \alpha - \beta c$. At the lower end of this range each has enough capacity to satisfy demand at $p_1 = p_2 = c$ and at the upper end of the range each has more than enough capacity to supply any conceivable demand he can face when he sets his price equal to cost. (When the products are perfect substitutes $k = \alpha - \beta c$ is just enough capacity for this. When $\gamma \neq 0$ this amount of capacity is more than enough as can be seen from OF' in Figure 6.1.)

When $k_1 = k_2 = (\alpha - \beta c)/2$, as was noted by Edgeworth, no equilibrium will exist. This implies that the maxmin will be above the competitive equilibrium.

We assume that the second firm in trying to inflict as much damage as possible on the first will utilize all of its capacity. Hence, it will set a price such that it will just utilize all of its capacity on the assumption that the other acts to maximize its profits. Given the price of firm 1, the price charged by firm 2 will be

$$p_2 = \frac{2(\alpha - 2k_2) + \beta\gamma p_1}{\beta(2 + \gamma)}. \tag{6.24}$$

Substituting this value for p_2 in the demand function of firm 1, the demand and profit functions for firm 1 are

$$q_1 = \frac{(1+\gamma)\alpha - \gamma k_2 - (1+\gamma)\beta p_1}{2+\gamma} \tag{6.25}$$

and

$$P_1 = q_1(p_1 - c). \tag{6.26}$$

Differentiating and solving for the optimal price for firm 1 yields

$$p_1 = \frac{(1+\gamma)(\alpha/\beta + c) - \gamma k_2/\beta}{2(1+\gamma)} \tag{6.27}$$

and

$$P_1 = \frac{\beta}{4(2+\gamma)(1+\gamma)}\left[(1+\gamma)\left(\frac{\alpha}{\beta} - c\right) - \frac{\gamma k_2}{\beta}\right]^2. \tag{6.28}$$

Using Eq. (6.27) in Eq. (6.24) we obtain

$$p_2 = \frac{(4+\gamma)(1+\gamma)\alpha/\beta + \gamma(1+\gamma)c - (8 + 8\gamma + \gamma^2)k_2/\beta}{2(1+\gamma)(2+\gamma)} \tag{6.29}$$

and

$$P_2 = k_2\left[\frac{(1+\gamma)(4+\gamma)[\alpha/\beta - c] - (8 + 8\gamma + \gamma^2)k_2/\beta}{2(1+\gamma)(2+\gamma)}\right]. \tag{6.30}$$

In the special case when $k_2 = (\alpha - \beta c)/2$ the prices and profits are respectively:

$$p_1 = \frac{(2+\gamma)\alpha/\beta + (2+3\gamma)c}{4(1+\gamma)}, \tag{6.31}$$

$$p_2 = \frac{\gamma\alpha/\beta + (4+3\gamma)c}{4(1+\gamma)}, \tag{6.32}$$

$$P_1 = \frac{(2+\gamma)\beta}{16(1+\gamma)}\left[\frac{\alpha}{\beta} - c\right]^2, \tag{6.33}$$

and

$$P_2 = \frac{\beta\gamma}{8(1+\gamma)}\left[\frac{\alpha}{\beta} - c\right]^2. \tag{6.34}$$

When $k_2 = \alpha - \beta c$ except when $\gamma = \infty$ this would require p_2 to be less than cost, as can be seen from Eq. (6.29).

Suppose that the second firm also holds an extremely conservative view of the first firm. Rather than trying to minimize the payoff of the first firm, the second firm believes that it must defend itself against the hostile behavior of the first firm. Thus it acts to select $\max_{p_2} \min_{p_1} P_2$. Given both firms behaving in this way, then

$$p_1 = p_2 = \frac{(1+\gamma)(\alpha/\beta + c) - \gamma k_2/\beta}{2(1+\gamma)} \tag{6.35}$$

and the profits are now higher to each because of their joint conservatism. In general, profits are

$$P_1 - P_2 = \frac{\beta}{8}\left[\left(\frac{\alpha}{\beta} - c\right)^2 - \frac{\gamma^2(k_2/\beta)^2}{(1+\gamma)^2}\right]. \tag{6.36}$$

Solving for the case in which $k_1 = k_2 = (\alpha - \beta c)/2$ we obtain price as in Eq. (6.31) and profits of

$$P_1 = P_2 = \frac{(2+3\gamma)(2+\gamma)}{32(1+\gamma)^2}\left(\frac{\alpha}{\beta} - c\right)^2. \tag{6.37}$$

It is of interest to note that the profit is always lower than that made at the quantity noncooperative equilibrium.

When we compare the maxmin solution with the price noncooperative equilibrium, profits can be higher or lower as a function of γ and k. Specifically, if $k_1 = k_2 = (\alpha - \beta c)/2$ and $\gamma > 2$, the maxmin solution will yield higher profits than the price equilibrium. In other words, extreme conservatism and distrust of each other by both may yield a result that is more jointly optimal than the noncooperative equilibrium.

If both firms have considerable excess capacity each is always in a position to reduce the profits of the other to zero. Hence $p_1 = p_2 = c$ and $P_1 = P_2 = 0$ are the prices and profits at an equilibrium that satisfies either Eq. (6.30) or Eq. (6.31). This coincides with the competitive equilibrium.

In the quantity or production model of competition the rule of maxmin behavior is simple. A firm on the defensive must assume that the other will produce up to capacity or to the level required to flood the market at cost, whichever is smaller. Thus, when the capacity of the other is $(\alpha - \beta c)/2$ it will produce $(\alpha - \beta c)/4$; hence if both fear each other they will jointly maximize in this instance!

Maximization of Market or Profit Share

It has been suggested on occasion that an oligopolistic firm strives to capture as large a market share as possible. This may be an appealing hypothesis but upon analysis it turns out to lack definition unless it is appropriately qualified. What price market share? The pursuit of market share under certain circumstances may spell disaster. Under other circumstances it may be closely correlated with profits, growth, and long-term success. The goal of maximization of market share calls for an explanation in terms of the dynamics of the market, or in terms of internal bureaucratic needs of the firm.

We must distinguish between two somewhat different approaches. A highly competitive firm may try to maximize its share of the profits of the industry. This may be described mathematically as

$$\max \frac{P_i}{\sum_{j=1}^{2} P_j}, \qquad \text{for} \quad i = 1, 2. \tag{6.38}$$

In the symmetric game for either the price or quantity, the profit share maximization leads to the competitive equilibrium when each has sufficient capacity!

A firm that is growth conscious may decide to accept a level of profit as a boundary condition that must be met to satisfy stockholders. Having met the boundary condition, the firm may aim at maximum expansion. If all firms follow this route, it can be expressed as

$$\max p_i q_i, \qquad \text{for} \quad i = 1,2 \qquad \text{subject to} \qquad P_i = (p_i - c)q_i \geq a_i, \tag{6.39}$$

where a_i is the profit level to be attained by the ith firm. To fully define this behavior the firms must make assumptions about each other. If each assumes that the other will remain a sitting duck, then each can decide what to do even though they might jointly fail to meet their targets.

Another version of maximize market share may call for

$$\max \frac{p_i q_i}{\sum_{j=1}^{2} p_j q_j}. \tag{6.40}$$

Usually when the phrase market share is used, the reference is to share of sales or gross revenues, hence market share maximization in this sense is reflected by Eq. (6.40).

Kinked Oligopoly Demand and Behavior

There has been a considerable literature, including the writings of Sweezy (1939), Stigler (1947), Krelle (1961), Sylos-Labini (1962), and others (Bronfenbrenner, 1940; Simon, 1969; Primeaux and Bomball, 1974) on the kinked oligopoly demand curve. This construct amounts to drawing a behaviorally dynamic contingent demand. Figure 6.7 illustrates the difference between this curve and the type of contingent demand considered here.

The curves DD' and dd' are the same as those found in Figure 6.4. The demand faced by firm 1 is dd' based on the assumption that firm 2 maintains its price at c and supplies whatever is demanded of it. Thus, the demand on firm 1 is *contingent* on firm 2 not changing price. From empirical

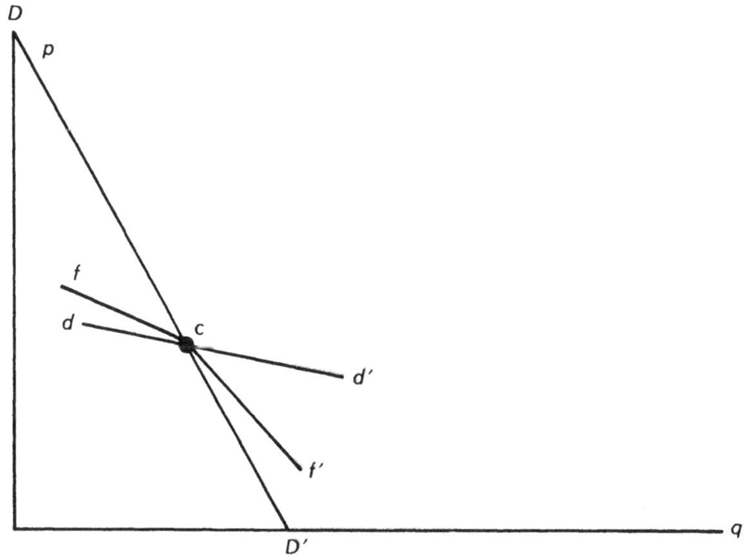

Figure 6.7. The kinked oligopoly demand.

information we could assume that if firm 1 moved its price up, firm 2 might react by raising its price not as much, but enough to create the expected demand curve *cf*. If, on the other hand, firm 1 were to cut price, firm 2 would react to bring about *cf'* as the contingent demand.

The problem of selecting which of the myriad conjectured behaviors can reasonably be used to predict outcome in a specific market is an empirical one that depends on detailed understanding of the level of communication, details of technology, time lags in organizations, and the many other factors that lend form to process and the dynamics of the market.

Those who wish to use kinked oligopoly demand curves to answer questions about competition in the steel or automobile industries would be well advised to know a great deal about the specific industries. Here our ambition is more limited. We are concerned with separating investigations of structure and behavior. Our first aim is to obtain a relatively exhaustive analysis of market structure, although this entails stressing static solutions. This is done as a preliminary to the eventual construction of a dynamic theory of oligopoly. If one wishes to pursue investigation of the kinked oligopoly demand curves, they can be considered special instances of the general set of oligopolistic reaction functions of the variety discussed by Stackelberg (1952), J. W. Friedman (1977), and others.

Reaction functions are related to but different from the historical strategies noted in Chapter 4. A reaction function, if treated in a purely automatic behavioral manner, does not really catch the full flavor of a strategic target. The kinked oligopoly demand discussion does not illuminate this important point. The mere writing down of difference or differential equations for the projected dynamics of all participants leads quickly to many cases of mutual idiocy, as is exemplified in models of conflict, such as those of Richardson (1960), applied to war.

An historical strategy as the example in Chapter 4 shows is a *threat* which is communicated, which may or may not be believed, and which may or may not be acted upon. A consideration of such threats leads to formal models, such as the Nash (1953) and Harsanyi (1963) cooperative games, and at a more formal level, the Selten (1973) cartel analysis. At a less formal level of analysis, this appears to be the basis for the Chamberlin (1962) small-group analysis and the quasicooperation discussed by Fellner (1949).

Product Differentiation or Variation

Product differentiation has been discussed here, but much of oligopolistic competition involves changing the level of differentiation among products, or varying products. The models presented here do not deal with this problem. Reflecting this new set of strategic opportunities would require γ as a variable. This possibility is discussed in Chapter 10.

A Simple Duopoly Game

Preliminaries and Parameters

In this chapter a duopolistic market was described and analyzed. In keeping with our desire to illustrate the theory and to enable the reader to calculate and to experiment with an actual game with ease, some parameters have been selected and solutions calculated using several different solution concepts.

Given the symmetric quadratic payoffs in the duopolistic market, the four key parameters are α, β, γ, and c. The capacity limitation k_i is also of importance. In Table 6.1 we calculate solutions for the following values:

The number of firms	$n = 2$
Market size	$\alpha = 2{,}000{,}000$
Slope of industry demand	$\beta = 2{,}500$
Substitutability parameter	$\gamma = 1, 3, 9,$ or 100
Average unit cost of production	$c = 200$
Capacity of each firm	$k = \alpha - \beta c = 1{,}500{,}000$

The α and β are selected to give the demand curve the appearance of one associated with a mass-produced item such as a major consumer durable. The variable α is the saturation level of consumption for the good if it were free, a state hardly to be expected for a consumer durable;

Table 6.1 Some duopoly game solutions.

	$\gamma = 1$			$\gamma = 3$			$\gamma = 9$			$\gamma = 100$		
SOLUTION	$P_i \times 10^6$	p_i	$q_i \times 10^3$	$P_i \times 10^6$	p_i	$q_i \times 10^3$	$P_i \times 10^6$	p_i	$q_i \times 10^3$	$P_i \times 10^6$	p_i	$q_i \times 10^3$
Individual monopolist[a]	150.00	500.00	500.00	180.00	500.00	600.00	204.54	500.00	681.81	222.79	500.00	742.65
Joint maximization	112.50	500.00	375.00	112.50	500.00	375.00	112.50	500.00	375.00	112.50	500.00	375.00
Quantity noncooperative equilibrium	110.20	457.14	428.58	106.09	430.77	461.38	103.02	412.90	483.87	100.33	401.32	498.36
Price noncooperative equilibrium	108.00	440.00	450.00	91.87	371.43	535.71	58.58	292.30	634.62	8.49	211.54	735.58
Competitive equilibrium	0	200.00	750.00	0	200.00	750.00	0	200.00	750.00	0	200.00	750.00

[a] From Eq. (6.11) and $P_i = q_i(p_i - c)$ we can calculate the monopolist's actions immediately: $\max_{q_i} P_i = q_i[\alpha/\beta - c - 1/\beta(2 + \gamma)/(1 + \gamma)q_i]$, which gives $q_i = [(\alpha - \beta c)/2][(1 + \gamma)/(2 + \gamma)]$ and $p_i = (V + c)/2$.

β controls the slope of *DF* in Figure 6.4 and γ controls the slope of *dd'* in Figure 6.4. The four levels of γ chosen correspond to isolated products, moderately close substitutes, close substitutes, and extremely close substitutes. The average cost c indicates the type of commodity. Is it a television set, a hi-fi, a dishwasher, or a stove? The capacity of the firms has been chosen to exceed the amount needed for a pure strategy price noncooperative equilibrium as described in Chapter 5.

In the actual playing of a game based on these parameters far more information must be supplied. That includes inventory costs, overheads, initial stocks, money costs, and costs of changing production levels and buying new capacity. For now, we leave these aside and turn to "solving" the market given the six basic parameters characterizing the symmetric duopoly.

SOLUTIONS

Table 6.1 gives the profits, prices, and productions associated with several of the more important solution concepts discussed in this chapter. They are each solved for four values of γ as given in the previous section.

The individual monopolist solution refers to the situation in which one firm inherits all of the market. When $\gamma = 0$ this inheritance does the monopolist no good as he is already isolated from his competitors' market. If $\gamma = \infty$, the individual monopolist would make precisely twice the amount obtained by a single firm if they split the proceeds of joint maximization, as the products would be perfect substitutes.

In the symmetric market with two firms the competitive equilibrium is also the maxmin-the-difference solution as well as the beat-the-average solution. If there is plenty of capacity the price noncooperative equilibrium solution also coincides with the competitive equilibrium.

Why bother with all of these different solutions? A key problem in the control of industry is to decide whether actions are to be directed at control of structure or behavior, and the role of intent. Depending upon the structure many different intents may result in the same behavior.

7 | Oligopoly with Product Differentiation

IN THE DUOPOLISTIC MARKET, representation of the preference of the customers led to demand functions of the form

$$d_i = \tfrac{1}{2}(\alpha - \beta[p_i + (p_i - \bar{p})]) \qquad (7.1)$$

for each player, where α, β, γ have the same meaning as in Eq. (6.1) and $\bar{p}$ is the average of all prices. For the n-firm symmetric market, the demand is modified to

$$d_i = \frac{1}{n}(\alpha - \beta[p_i + (p_i - \bar{p})]). \qquad (7.2)$$

By introducing $1/n$ into the demand relationship, a class of market can be constructed that is of value in the study of the effect of numbers on the nature of competition. We begin by examining a market of given size. When $n = 1$, there is a monopolist servicing the market. The term containing γ drops out as there is only one supplier. When $n = 2$ the monopolist has been replaced by duopolists, each of whom obtains a demand of one half the size of the market held by the monopolist when each duopolist charges the same price and that price is the amount charged by the monopolist. In a similar manner when there are n firms in the market, each obtains a $1/n$th market share when the same price is charged.

Since the firms are not selling identical products but are selling symmetrically related substitutes it is not strictly correct to draw individual contingent demands on a two-dimensional diagram as one would be adding "oranges and apples." Given the symmetry, however, Figure 7.1 may be used with careful interpretation. If a monopolist charges $p = 0$ he will sell $0F$. The demand faced by the monopolist is DF. An individual duopolist faces DD_2 when all charge the same price. Similarly for four firms the individual firm faces DD_4 *if* all firms are charging the same price. If all firms but one charge the same price, its demand (in the four-firm case) will vary as the firm varies its price along dd' in the neighborhood of s. The slope of the dd' curve will be $(\beta/n)[1 + (n - 1/n)\gamma]$. When several firms are charging different prices the complete algorithm described in Appendix B must be used to describe the general contingent demand curves that exist.

Solutions for a Symmetric Oligopolistic Market with Product Differentiation

Joint Maximization and the Efficient Solutions

If the firms have constant and equal average costs and the market size is the same for an oligopoly with any number of competitors (we are

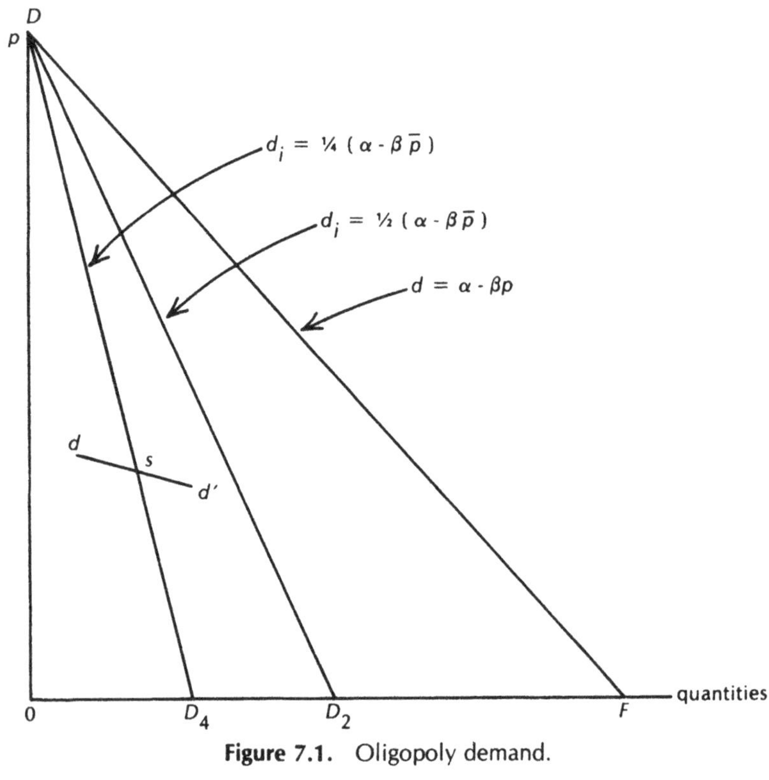

Figure 7.1. Oligopoly demand.

chopping up the firms), the prices charged by each at the joint maximum and at efficient production will not change regardless of the number of competitors involved. Thus the joint maximum price and total industry profit are

$$p = \frac{\alpha + \beta c}{2\beta} \tag{7.3}$$

and

$$P = \frac{1}{\beta}\left(\frac{\alpha - \beta c}{2}\right)^2 \tag{7.4}$$

and individual sales are

$$q_i = \frac{\alpha - \beta c}{2n}. \tag{7.5}$$

The price, outputs, and profits at the efficient point are

$$p_i = c, \quad q_i = \frac{\alpha - \beta c}{n}, \quad \text{and} \quad P_i = 0. \tag{7.6}$$

PRICE AND THE QUANTITY NONCOOPERATIVE EQUILIBRIA

The price or the Chamberlinian large-group noncooperative equilibrium can be calculated from

$$\frac{\partial P_i}{\partial P_i} = 0 \qquad \text{for} \quad i = 1, \ldots, n,$$

where

$$P_i = \left[\frac{1}{n}(p_i - c)(\alpha - \beta[p_i + \gamma(p_i - \bar{p})])\right], \tag{7.7}$$

which yields

$$p_i = \frac{\alpha/\beta + c[1 + (n-1)/n]\gamma}{2 + [(n-1)/n]\gamma}. \tag{7.8}$$

The quantity sold by each firm and its profits are

$$q_i = \left(\frac{1 + [(n-1)/n]\gamma}{2 + [(n-1)/n]\gamma}\right)\left(\frac{\alpha - \beta c}{n}\right)$$

and

$$P_i = \frac{1}{\beta n}\left(\frac{\alpha - \beta c}{2 + [(n-1)/n]\gamma}\right)^2. \tag{7.9}$$

Similarly we may examine the quantity or production strategy equilibrium using the inverse form of Eq. (7.2) which expresses price as a function of quantities:

$$p_i = \frac{\alpha}{\beta} - \frac{\gamma}{\beta(1+\gamma)} \sum_{j \neq i} q_j - \frac{n+\gamma}{\beta(1+\gamma)} q_i. \tag{7.10}$$

The payoff function for i is

$$P_i = q_i(p_i - c), \tag{7.11}$$

where p_i is defined in Eq. (7.10). Solving for the first-order optimization conditions we obtain

$$q_i = \frac{(1+\gamma)(\alpha - \beta c)}{n(2+\gamma)+\gamma}. \tag{7.12}$$

The price charged by each firm is

$$p_i = \frac{(1+\gamma/\eta)\alpha/\beta + (1+\gamma)c}{2+(1+1/n)\gamma}. \tag{7.13}$$

As n increases the price in Eq. (7.8) or Eq. (7.13) does not approach that of the efficient point or competitive equilibrium. The meaning of this model is that although many firms with substitutes are crowding into a limited market the substitutes remain apart. If we make α, β, and γ all functions of n so that $\alpha_n = n\alpha$, $\beta_n = n\beta$, and $\gamma_n = n\gamma$, we can see from Eqs. (7.8) and (7.13) that as $n \to \infty$, $p_i \to c$. This means that as the number of competitors increases, the substitutes come closer together.

The introduction of more than two firms by no means guarantees the existence of a price equilibrium in the market. If there is insufficient excess capacity the instability shown by Edgeworth may still exist. Shapley and Shubik (1969b) have shown that the amount of excess capacity needed by each firm in a symmetric market varies as the number of firms. The explicit formula is given in Eq. (7.14) below. If the firms have less than this amount of capacity no price equilibrium will exist.

$$k_i = \frac{n^2(\alpha - \beta c)(1+\gamma)}{2(n-1)(n\gamma + n - 2)} \times \left[1 - \frac{1}{n} \frac{\sqrt{[n(1+\gamma)+2(n-1)][n(n-1)(1+\gamma)+2]\left[\dfrac{2}{\beta(1+\gamma)}\right]}}{2(n+1)+n(n-1)(1+\gamma)}\right] \tag{7.14}$$

An interpretation of mixed strategies and price instability is given in Chapter 8, when capacity constraints are considered in greater depth.

Pareto Optimality and the Threat Curve

We obtain a straightforward generalization of the Jacobian condition given in Eq. (5.22) to define the Pareto optimal surface and the threat curve. The interpretation of the threat possibilities is, however, somewhat different; this is noted below.

In defining the Pareto optimal surface the definition for two or many firms is the same. A point is on the surface when it is not possible to improve the payoff to one individual without making the payoff worse to another. When we define the threat curve, however, we must specify how many firms participate in making or reacting to threats. If we consider the n firms split into two groups consisting of r firms that threaten and $n - r$ firms that protect themselves, there would be $n - 1$ different threat possibilities in a symmetric market as the group of threateners varies from 1 to $n - 1$.

The amount of damage that an individual firm can do to its competitors in a symmetric market will attenuate rapidly as the number of firms increases. This would not be true if an individual firm had enormous overcapacity. It is worth stressing again that capacity is basically a dynamic concept related to investment and overheads. Capacity is often difficult to define especially with multiproduct, multiplant firms, however particularly for one-product firms, excess capacities of over several hundred percent would be a sign of a sick industry indeed!

Mathematically it is easy to define the conditions for the threats of any r firms against the other $n - r$ (Shubik, 1959b). However, if none of the firms has excess capacity but each has enough for his output at competitive equilibrium, all threats are the same and amount to producing the competitive equilibrium output.

Beat-the-Average and Maximize Profit Share Solutions

In Chapter 5 it was noted that two firms might try to maximize the difference between their profits. This was expressed as

$$\max \min (P_1 - P_2). \tag{7.15}$$

The generalization of this solution concept to a symmetric market with many competitors is straightforward. We may assume that the individual tries to maximize the difference between his score and the average, trying to beat-the-average. This may be expressed as

$$\max \left(P_i - \frac{1}{n-1} \sum_{j \neq i}^{n} P_j\right). \tag{7.16}$$

For $n = 2$, Eq. (7.16) is the equivalent of Eq. (7.15). Furthermore for any n, the symmetric beat-the-average solution is the same as the competitive equilibrium.

Apart from being an n-person generalization of the maxmin-the-difference solution, the beat-the-average solution has another important interpretation. Very often when situations are complex and good measures are hard to obtain, the firm itself or market analysts judge it by its performance relative to other firms. This implicitly introduces a level of competition far greater than is commonly realized.

Whenever a market game or any other competitive game is run for experimental purposes it is usually highly desirable to check the behavior of the individuals against a beat-the-average hypothesis. Often no matter how carefully the players have been instructed concerning their goals, they turn the game into one of intense competition. This seems to be especially true of some people when playing an artificial player. The ego involvement in being challenged by a computer program may bring out an extremely competitive attitude.

In Chapter 9 the maximization of profit share is shown to be the appropriate generalization for a nonsymmetric market. At this point, however, the reader should check to see that in a symmetric market, maximize profit shares and beat-the-average solutions are equivalent.

How Many Is Many?

A Measure of Many

How many is many? In the context of the symmetric market we are in a position to offer an answer to this question. We define a market as having many competitors when the number of competitors is sufficiently large so that the noncooperative equilibrium prices of the firms are within $g\%$ of the prices at efficient production, where the total range is given by the difference between prices at the joint maximum and at the efficient production level.[1]

For our interest in the concept of many in a market, the question we want to ask is given a specific market with n_1 identical firms, how does it differ from another market with $n_2 > n_1$ identical smaller firms? An equivalent way of asking this question is to add firms of the same size and enlarge the market. Thus, for purposes of comparison, we consider $\alpha_n = n\alpha$, $\beta_n = n\beta$, and $\gamma_n = n\gamma$. Using this transformation we note that the equations for price in (7.3) and (7.6) are unchanged, but the noncooperative equilibrium price becomes

$$p_i = \frac{\alpha/\beta + [1 + (n-1)\gamma]c}{2 + (n-1)\gamma}. \qquad (7.17)$$

1. One might wish to make the definition in terms of profits or the range of Consumer's Surplus; however, prices may be more easily measured. Lerner (1934) has suggested measuring monopoly power as an index of (price-marginal cost)/price. Measures and descriptions of industrial concentration are given by the Lorenz curve and Gini Coefficient. The Herfindal index provides a measure composed of the sum of the squares of market share. (See Adelman, 1969.)

The criterion to decide when the (price) noncooperative equilibrium[2] lies within $g\%$ of the efficient point is given by

$$c + g\left(\frac{\alpha/\beta - c}{2}\right) \geq \frac{\alpha/\beta + [1 + (n - 1)\gamma]c}{2 + (n - 1)\gamma}.$$

We obtain

$$n = \frac{2 - g(2 - \gamma)}{g\gamma}. \tag{7.18}$$

For example, suppose $g = 0.05$ and $\gamma = 9$, we have

$$n \geq \frac{1.45}{.45}.$$

Hence, four firms are enough for the price noncooperative equilibrium to be within 5% of the competitive equilibrium.

What constitutes many in a market is undoubtedly a function of the levels of communication among the firms, accepted laws and customs of the society, and a host of behavioral considerations. For our analysis the size of many is related to the solution concept chosen. Recently Selten (1973) developed a cartel behavior solution in which many is 6 or more. In a single market many is most frequently between 2 and 20 if formal collusion is forbidden.

We can see from Eqs. (7.8) and (7.13) that the speed of convergence is $1/n$. This is a property of the linear demand and constant marginal cost of production. No general work has been done to depict the rate of convergence toward an efficient market for general conditions and a noncooperative solution. However some work has been done under extremely general conditions to study the convergence of the core in a general equilibrium setting. Shapley has shown that it is possible to construct examples with an arbitrarily slow speed of convergence; Debreu (1975) has shown that (under a somewhat technical definition of measure) the Shapley (1975) examples will be rare.

From the viewpoint of oligopoly theory and application it is conjectured here that for most market structures and solutions competition is approached at a rate of $1/n$ if payoffs are differentiable.

A Measure of Implicit Collusion

It has been shown that under certain circumstances many different solution concepts would predict the same outcome for a competitive situa-

2. Using Eq. (7.13) rather than (7.8) a somewhat different criterion for "many" under quantity equilibrium can be derived. It will be seen that many calls for larger numbers with quantity competition.

tion. Suppose that we observe an outcome in a symmetric market. Is there any way in which we can at least obtain an approximate measure of the degree of cooperation or collusion present in that market? By using the solution suggested by Edgeworth we could obtain a measure of cooperation by estimating the parameters θ_i:

$$\prod_i = P_i + \theta_i \sum_{j \neq i} P_j. \tag{7.19}$$

The meaning of the use of this measure can be seen from Figure 7.2. Comparing two markets, one with two firms and the other with three, we would observe the points C_2 and C_3 if our estimates indicated that $\theta_i = 0$. The profits made with three firms in the market are far less than those made with two firms even though the parameter measuring the level of cooperation has the same value. This enables us to say that while we might deem the intents or the degree of cooperation to be the same, the market structure is more conducive to price distortion with only two firms than with three firms. Suppose that there were three firms in the market and we observed the point B_3. The profit at this level is no higher than at C_2; however, we can measure the level of cooperation required in a three-firm market to obtain the outcome B_3 by a positive θ_i even though this same profit could be obtained in a duopolistic market with no collusive intent whatsoever.

If in an actual market the outcome B_3 were observed this might be sufficient reason to take action against the firms on the basis of intent or behavior rather than market structure.

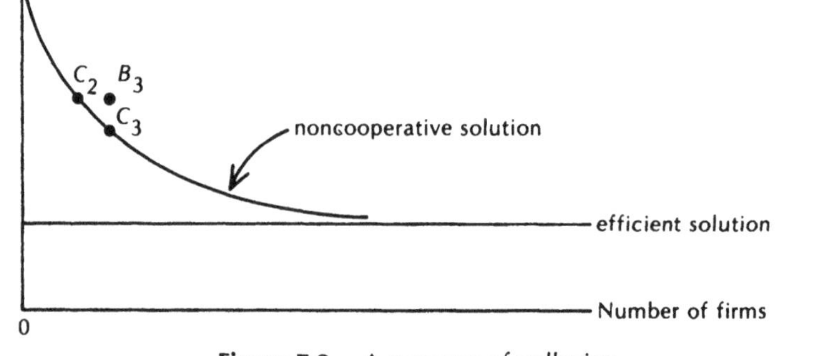

Figure 7.2. A measure of collusion.

Welfare, Efficiency, and Symmetry

The assumption of symmetry is a considerable simplification as will be seen in Chapter 9. Once it is abandoned, not only does the mathematical analysis become considerably more difficult, but certain meaningful and reasonable concepts with symmetry are no longer valid and must be replaced. Even with symmetry, numbers and differentiation of product cause problems in the measurement of welfare and efficiency. The difficulties are more conceptual than mathematical.

With all firms selling the same product the measurement of consumer surplus is straightforward. Furthermore, with symmetry, it is useful to analyze producers of identical items having the same costs since there is a natural unit of production to evaluate. We can deduce that efficiency calls for the lowest cost producer to satisfy the market. If this producer has sufficient capacity, or if all firms have the same costs, it does not matter who produces for efficiency is not affected. With differentiated products this is not the case. If products are differentiated while still reasonably similar, costs are no longer sufficient to determine efficiency. Consumer welfare is sensitive to variety, but how much variety? The distance M_1J_1 in Figure 7.3 gives an indication, where J_1 is the profit of the monopolist selling a differentiated product, and M_1 is profit if the products were perfect substitutes.

In the utility function given in Eq. (6.1) the oligopolistic analogue that yields the demand given by Eq. (7.2) calls for an increasing welcoming of diversity—the more products there are the more variance is welcomed. After a while this model of the consumer does not appear to be completely reasonable.

Cross-Elasticity, Market Structure, and Strategy

Elasticity and cross-elasticity of demand are measures of market sensitivity to change in price. In particular, when dealing with oligopolistic markets, three concepts must be sorted out. We may wish to know how the overall market for automobiles or cigarettes is affected if all prices move in concert. This gives us an overview of the sensitivity of the whole market with regard to price. We may wish to know what happens to the revenues of an individual firm as it changes its price, *all other prices remaining fixed,* and we may want to know how a change of price by one firm affects the revenues of any other firm, all other prices remaining fixed. All three are measures of the strategic interlinkage among the firms in the market.

All of the elasticity measures are of the form

$$e = \frac{p\partial q}{q\partial p} = \frac{\partial \log q}{\partial \log p}. \tag{7.21}$$

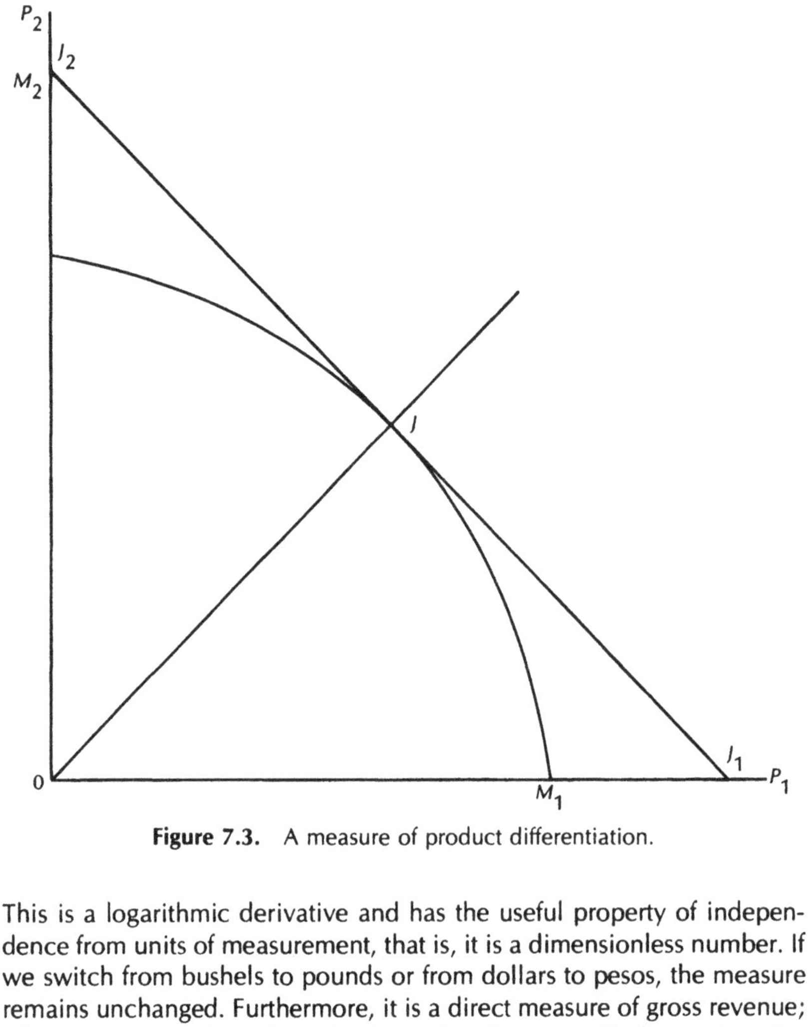

Figure 7.3. A measure of product differentiation.

This is a logarithmic derivative and has the useful property of independence from units of measurement, that is, it is a dimensionless number. If we switch from bushels to pounds or from dollars to pesos, the measure remains unchanged. Furthermore, it is a direct measure of gross revenue; when it assumes the value of 1 it implies that a small change in price leaves gross revenue unchanged. This is illustrated in Figure 7.4. At the point E gross revenue is given by $OHEG$. Figure 7.4 represents the market for four symmetric firms. DF is the overall market for the generic product without brand distinction. DD_4 represents the demand faced by a single firm on the assumption that all prices move together. Based on the assumption that the prices of all other firms remain constant while the price of a single firm changes, dd' represents the demand faced by the single firm. The isorevenue curve R is marked and the point E lies on it.

Elasticity measures are relevant to a particular point on a demand or contingent demand curve. An elasticity measure gives information about revenue sensitivity only in the neighborhood of specific array of prices and sales. When we talk about an oligopolistic market and wish to describe the competitiveness of the market we are usually referring to the market with the firms in some particular state. We choose to consider elasticity when the firms are in a state of price competition noncooperative equilibrium. In Figure 7.4 we measure elasticities at the points *T* and *E*.

At the point *T* the overall market elasticity for all firms considered as an aggregate is

$$e = \frac{\bar{p}dq}{qd\bar{p}} = \frac{\bar{p}d(\alpha - \beta\bar{p})}{qd\bar{p}} = \frac{\beta p}{q}, \tag{7.22}$$

where $\bar{p}$ and q are the average price and total sales at noncooperative equilibrium and e is defined as the *negative log derivative*. Hence, our elasticity measures are positive with a demand curve sloping down to the right.

At the point *E* we may define the two other elasticity measures. First,

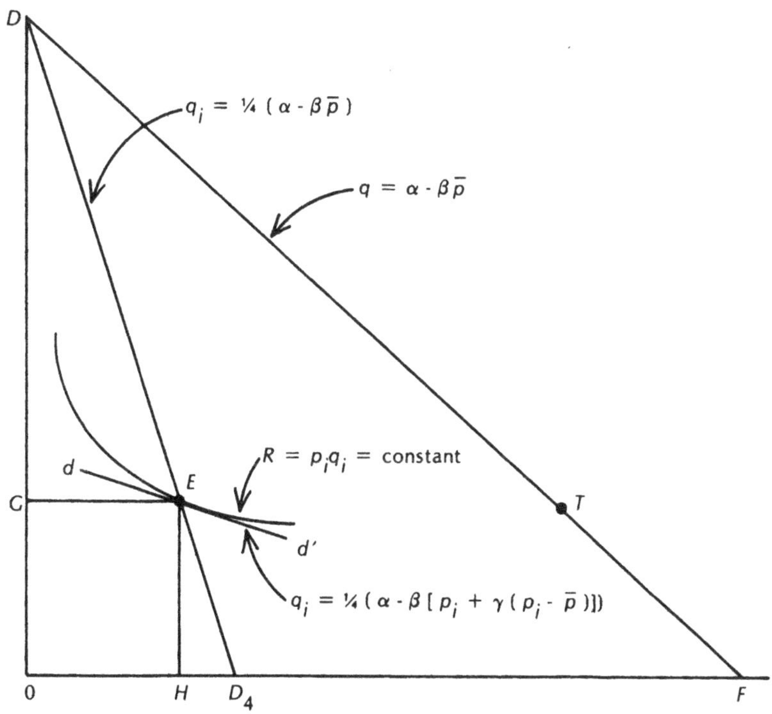

Figure 7.4. Oligopolistic revenues.

using Bishop's notation we define the own elasticity of firm i as

$$E_{ii} = \frac{p_i}{q_i}\left(\frac{\partial q_i}{\partial p_i}\right) = -\frac{\beta}{n}\left(1 + \gamma\frac{(n-1)}{n}\right)\frac{p_i}{q_i}. \tag{7.23}$$

Now we see the relationship between the elasticity of the firm and the noncooperative price equilibrium. For simplicity let us suppose that the costs of production are zero. The price noncooperative equilibrium in a symmetric market will be at the point E on the line DD_4 where $E_{ii} = -1$, that is, the elasticity at E along dd' is unity. We can check this by substituting in the values of p_i and q_i into Eq. (7.23) using the condition that $c = 0$.

In the symmetric market, $p_i = \bar{p}$ and $q = nq_i$ at equilibrium. We can use these conditions in Eq. (7.22) to evaluate e at the point T.

The last measure to be defined is the cross-elasticity, or the effect on the revenues of firm j caused by the price change of firm i:

$$E_{ji} = \frac{p_i}{q_j}\left(\frac{\partial q_j}{\partial p_i}\right) = \frac{\beta\gamma}{n^2}\left(\frac{p_i}{q_j}\right). \tag{7.24}$$

Bishop has used the index $-E_{ii}/E_{ji}$ to measure or characterize the oligopolistic structure of a market:

$$-\frac{E_{ii}}{E_{ji}} = \frac{(1 + [(n-1)/n]\gamma)}{\gamma/n}. \tag{7.25}$$

With respect to this particular market structure we may check and interpret $\gamma = 0$, $\gamma \to \infty$, $n = 1, 2, 3$, and $n \to \infty$. From Table 7.1 we can interpret for $\gamma = 0$ that we have isolated markets. There is no interaction among the firms. For $\gamma \to \infty$ the values are formally correct but not particularly useful. When $\gamma \to \infty$ the contingent demand dd' becomes horizontal implying that products are extremely close substitutes. The values merely indicate that if there are $n - 1$ other firms in a market they lose or gain equally from the change of price of the remaining firm.

The entry for an unspecified γ shows how the index increases in a linear manner with the growth of the number of competitors. This reflects the loss of monopoly power caused by the presence of additional substitutes.

A host of suggestions have been made for a taxonomy of market structures. We do not propose to enter into a detailed discussion here. The reader is referred to some of the literature (Bishop, 1952; Machlup, 1952a; Pfouts and Ferguson, 1959). We merely note that the three measures e_1, E_{ii}, and E_{ji} are useful in describing the strategic interlinkage in a market caused by price competition. Price competition is by no means the only form of oligopolistic competition, but it is usually important and

Table 7.1 Own and cross elasticities.

VALUES OF $(-E_{ii}/E_{ij})$	γ		
	0	γ	∞
1	∞	$1/\gamma$	0
2	∞	$\frac{2+\gamma}{\gamma}$	1
3	∞	$\frac{3+2\gamma}{\gamma}$	2
⋮	⋮	⋮	⋮
n	∞	$\frac{n+(n-1)\gamma}{\gamma}$	$(n-1)$
⋮	⋮	⋮	⋮
∞	∞	∞	∞

hence the elasticity measures can be used as a shorthand to describe the structure of market sensitivity.

In this section we have stressed the importance of sorting out questions of intent, behavior, and structure. We believe that it is unreasonable to impute intent to collude to firms whose behavior can be described by a noncooperative theory. Yet at the same time market structures caused by technological conditions and fewness may make the noncooperative outcome still unacceptable. In this case structure may require change.

A Simple Oligopoly Game

We now extend the model presented in Chapter 5 to the market with many firms. It is noted that it is somewhat easier (though logically equivalent) to compare markets with different numbers of symmetric firms by expanding the market and keeping the firms the same size, rather than by cutting up the firms. This has the convenient feature that for markets of any size the price, production level, and profits of the individual firm assuming either joint maximization or competitive equilibrium remain the same.

The following parameters are used:

$$\alpha = 1{,}000{,}000n \quad \text{for} \quad n = 1, 2, 3, 4, 5, 7, 10, 20$$
$$\beta = 1{,}250n$$
$$\gamma = 1, 3, \text{ or } 9$$
$$c = 200.$$

Table 7.2 Price outcomes of oligopoly games.

Solution	Oligopoly price							
	$n = 1$	2	3	4	5	7	10	20
Joint maximum	500	500	500	500	500	500	500	500
Noncooperative equilibrium								
(*quantity*): $\gamma = 1$	500	457.14	440	430.77	425	418.18	412.90	506.56
$\gamma = 3$	500	430.77	400	382.61	371.43	357.89	347.17	333.98
$\gamma = 9$	500	412.90	371.43	347.17	331.25	311.63	295.80	275.98
Noncooperative equilibrium								
(*price*): $\gamma = 1$	500	440	425	418.18	414.29	410	406.90	403.39
$\gamma = 3$	500	371.43	350	341.18	336.36	331.25	327.66	323.71
$\gamma = 9$	500	292.30	275	268.57	265.22	261.76	259.41	256.87
Competitive equilibrium	200	200	200	200	200	200	200	200

Table 7.3 Profit outcomes of oligopoly games.

Solution	Oligopoly profit (per firm)							
	$n = 1$	2	3	4	5	7	10	20
Joint maximum	112.5	112.50	112.50	112.50	112.50	112.50	112.50	112.50
Noncooperative equilibrium								
(*quantity*): $\gamma = 1$	112.5	110.20	108.00	106.51	105.47	104.13	103.02	101.59
$\gamma = 3$	112.5	106.09	100	95.27	91.84	87.26	83.30	78.05
$\gamma = 9$	112.5	103.02	91.84	83.30	76.90	68.14	60.38	49.77
Noncooperative equilibrium								
(*price*): $\gamma = 1$	112.5	108	105.47	104.13	103.32	102.38	101.66	100.83
$\gamma = 3$	112.5	91.84	84.38	80.97	79.03	76.90	75.37	73.65
$\gamma = 9$	112.5	58.54	49.22	45.55	43.60	41.55	40.14	38.61
Competitive equilibrium	9	0	0	0	0	0	0	0

Table 7.4 Oligopoly market elasticity measures.

	Elasticity measures						
Solution	$n = 2$	3	4	5	7	10	20
Joint maximum							
e	1.667	1.667	1.667	1.667	1.667	1.667	1.667
$-E_{ii}$	2.604	5.000	5.417	5.667	5.952	6.167	6.417
E_{ji}	2.500	1.667	1.250	1.000	.714	.500	.250
Noncooperative equilibrium							
(*price*): e	.866	.777	.744	.725	.707	.694	.680
E_{ii}	2.167	2.330	2.417	2.467	2.524	2.567	2.617
E_{ji}	1.300	.777	.558	.435	.303	.208	.102
Competitive equilibrium							
e	.333	.333	.333	.333	.333	.333	.333
E_{ii}	.833	1.000	1.083	1.333	1.190	1.230	1.283
E_{ji}	.500	.333	.250	.200	.143	.100	.050

For $n = 2$, these parameters are identical to those in Chapter 5. Table 7.2 presents the prices for the different solutions in markets with varying numbers of players and the three values of γ. Table 7.3 gives the profit information for the individual firm under the same circumstances. Table 7.4 presents the elasticity measures e, E_{ii}, and E_{ji} in these markets evaluated at several of the different solutions.

8 | Capacity, Inventories, and Demand Fluctuations

THE LIMITED NATURE of the solutions to most static oligopoly models such as the Cournot and price duopoly models indicates that several steps closer to reality must be taken to construct a satisfactory theory of oligopoly. A step toward dynamics is undoubtedly called for. It is tempting to make a jump from the microstatics to a grand theory of dynamic oligopoly. This, to a certain extent, has been the tradition from Chamberlin (1962) onward. The kinked oligopoly curve approach, the models of Stackelberg (1952), and even the models of Fellner (1949), Brems (1951), Bain (1956), Baumol (1959), Sylos-Labini (1959), and Williamson (1975), attempt to give us a broad picture that can be used for policy guidance and insights into the control and development of industry.

The inductive leaps providing the appropriate reaction functions, the correct aggregation of financial variables, the key characterization of technology, the appropriate insights into the interface between public and private administrative bureaucracies are to be welcomed. Administrators and other decisionmakers cannot wait forever until theory is placed on solid ground. This being the case, the attempts of economists to sketch a broad approach to the dynamics of industry without becoming bogged down in too many institutional, technological, or financial details is highly desirable for the purpose of providing some form of immediate, thoughtful economic theorizing about the control of industry.

The approach adopted here, however, is somewhat different and less ambitious, or ambitious in a different direction. This approach is based on the belief that an adequate theory of markets and oligopolistic behavior calls for a peculiar mix of institutional, technological, and behavioral details with many variations of mathematical models. The blend yields a mathematical institutional economics in which the models are extremely ad hoc when applied to any particular firm or industry but the methods are general and should serve to identify the key variables that differ from industry to industry and provide a means for creating better hand-tailored models that are compatible with basic economic theory, management science, and operations research.

Rather than jump immediately to an intermix of assumptions concerning actions and reactions, the method followed here is to separate as much as possible assumptions concerning market and firm structure from behavior.

In particular we begin now to elaborate the simple models presented earlier by taking into account some of the implications of exogenous uncertainty in the market, production time lags and inventory problems, the influence of capacity limitations, and the possibility of stockout penalties. In some industries, these factors may be minor; in others they are critical. Many books (Whitin, 1953; Buchan and Koenigsberg, 1963; Wagner, 1969) and thousands of articles already exist in the literature of management science, operations research, and economics on production and inventory scheduling. No effort is made here to provide a coverage of this literature. However, the links between this work and oligopoly theory are hopefully suggested in the subsequent models.

Monopoly, Inventories, Overheads, and Uncertainty

In verbal and diagrammatic descriptions of the behavior of the single-product firm in a monopolistic or competitive market, average costs are often portrayed by a U-shaped curve with no particular upper bound and with little if any specification of the time period for which the costs are defined. In the microeconomic description of the firm for the study of mathematically formulated dynamic models greater specificity is needed than was called for by the earlier approaches. In particular, the meaning of capacity restrictions must be made specific and short or long-term costs and decisions must be specified in much the same way as the accountants describe the firm.

If we accept some ad hoc definition of the short-run decision period for the firm, be it three months or a year, the meaning of a short-term upper bound to capacity takes on a fairly straightforward and reasonable aspect. Presumably production requires plant and management. There is an upper bound to what can be produced with existing plant and management and in the space of the period under consideration it is not possible to increase either management or plant to be instantly effective. If plant and other overheads are required for production and they are not costless, then an overhead cost must be assigned.

Let k_t be capacity at time t; b is the cost of a unit of capacity; ρ is the cost of capital to the firm. The carrying cost of the plant is then ρbk_t on the money tied up. If g is the rate of depreciation, there is a charge (but not necessarily a cash flow) of gbk_t on depreciation. There are also administrative overheads.

Once we begin to be specific about time periods, several details concerning production must be noted. In particular, is all production ordered during period t available for sale during t? In this chapter we assume yes; a more reasonable assumption may be that half of current production (or some other fraction) is available with the remainder being added to inventories.

The importance of inventories depends on both the specific nature of the item and its demand. A few examples of the turnover rates governing the size of inventories are given in Table 8.1. At the least, inventory carrying costs equal the interest rate on the money invested. When we include packaging, storage, obsolescence, theft, and deterioration, inventory carrying may run anywhere from 5%–10% per annum, of manufactured costs to 30%, 40%, or more. To some extent, this is a matter of accounting and definition of the various inventory processes. There are raw goods, goods-in-process, and finished goods inventories. Limiting ourselves to finished goods, there are inventories held at the factory, those transported to and held at the distributor's warehouse, and those occupying expensive selling space on display in retail outlets.

An important problem in accounting is how to evaluate inventories held at the end of an accounting period. Custom has it as cost or market, whichever is less. This presumes that a well-defined market price exists, which is not always the case.

The one-period, or short-run maximization problem facing the firm involves pricing, production, and inventories. The longer-run problem calls

Table 8.1 Median inventory turnover for various industries.

Industry	Number of days
Retail groceries	23
Dairy products	14
Book publishing and printing	155
Retail automobiles	56
Retail luggage	104
Retail liquor	54

Source: Robert Morris Associates, *Annual Statement Studies,* 1969, pp. 89, 198, 199, 208, 209, 211.

for adjustment of capacity. Here we consider in a static framework the one-period model of duopolistic firms with the added complications of inventories, stockout costs, and exogenous uncertainty.

Before we investigate duopoly or oligopoly models it must be noted that the introduction of uncertainty poses new problems for the single firm as a monopolist or as a unit in a competitive market. This problem has been considered in detail both in one period and multistage versions (Veinott, 1966; Baron, 1970; Zabel, 1970; Leland 1972a,b). It is of importance to note that the analysis depends explicitly on the choice of price or quantity as control variables and on the fine details of sequencing of moves and information.

Price and Price-Quantity Duopoly: Mixed Strategies and Capacity

Actual oligopolistic competition is so fundamentally dynamic that our attempts to construct well-defined static models may lead to formally correct but empirically unsatisfactory models. This somewhat painstaking approach to examining many variations of simple models nevertheless has the advantage that the significance of different variables can be examined in relative isolation.

In Chapters 5, 6, and 7 the Edgeworth cycle was encountered. The idea of price behaving as a yoyo in a single period is not entirely reasonable. The model requires a dynamic setting. But even before an attempt is made to consider full dynamics a closer view of the cause of the instability is worthwhile.

Price Strategy Duopoly with Capacity Constraints

In particular the capacity of the firm to produce, the length of time of production, inventory carrying costs, and stockout losses all may play a role in the price strategy duopoly. If firms do not have excess plant, as has been seen in Chapters 6 and 7, the conditions for price instability are present. Even with excess capacity, production takes time and inventory costs are not zero. Instability may be caused by a very short-run lack of production although enough long-run capacity exists. The possibility of penalties for being out of stock, such as losing the customer next time, also modifies behavior. To demonstrate clearly the effect of a restriction of capacities or production we consider a starkly simple model that fits with those of Chapter 5.

Let the market demand be

$$q = a - p. \tag{8.1}$$

Capacities are given by k_1 and k_2. To begin, we assume $k_1 = k_2 = k$, that is, the firms have equal capacity. Production costs are zero. Furthermore,

$k \leq a$, that is, a firm has no more capacity than that needed to supply the entire market.

Suppose the firms charge prices p_i and p_j. We assume demand as follows:

$$q_i = \begin{cases} a - p_i & \text{if } \; p_i < p_j, \\ \tfrac{1}{2}(a - p_i) & \text{if } \; p_i = p_j, \\ a - k - p_i & \text{if } \; p_i > p_j. \end{cases} \tag{8.2}$$

The price and total output in the Cournot market are shown at point C in Figure 8.1. Point J gives the price and output if the firms act together as a monopolist.

Edgeworth introduced the possibility of limited capacity. The case for $k = 3a/4$ is illustrated in Figure 8.2. Suppose that one firm were charging $p_i = 0$; it can only sell up to capacity k. This would give the other firm a demand described by DD'. It would pay to seek monopoly profit against

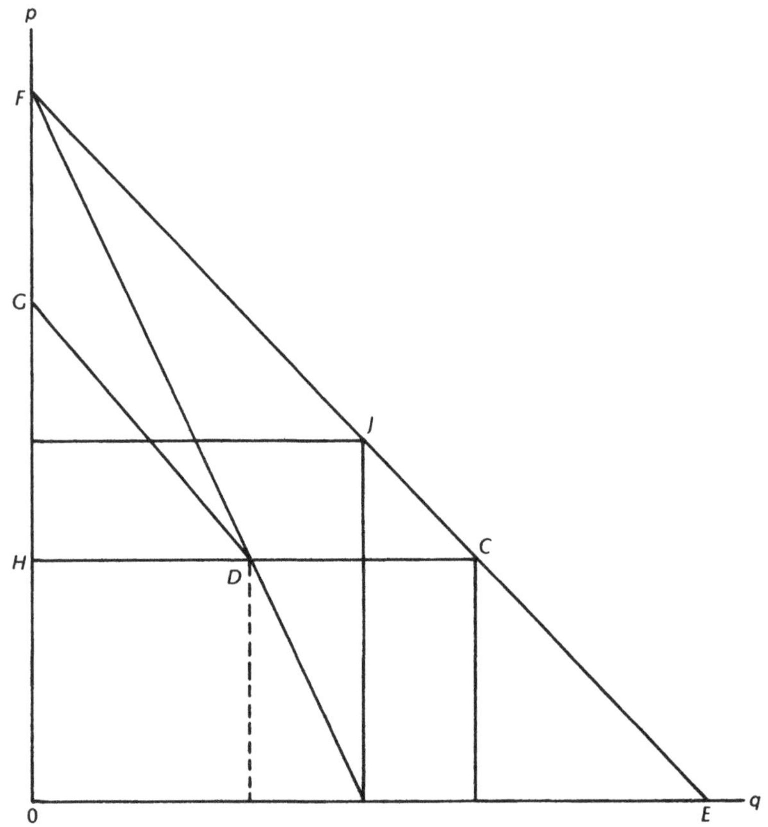

Figure 8.1. Duopoly with restricted capacity: case 1.

this contingent demand by raising price to M as shown in Figure 8.2. Subsequently, if we are willing to follow a loose dynamic argument, the other firm may raise its next price to just under that at M and a period of price-cutting may follow. Thus Edgeworth suggested that there was a range over which price might be expected to fluctuate. This range will depend upon the capacity k.

It is simple to solve for the upper and lower bounds of this range. Let the price at the bottom of the range be $\bar{p}$. Suppose that the price of one firm is indefinitely close to the bottom of the range and that the other firm has the choice of picking the price at the bottom of the range or raising its price to $\hat{p}$. It will be indifferent if profits are equal, that is, if

$$k\bar{p} = (a - k - p)p. \tag{8.3}$$

Fortunately in this simple example the profit to the high-priced firm depends only on its price and the capacity of the other; hence we know that

$$\hat{p} = \frac{a - k}{2}. \tag{8.4}$$

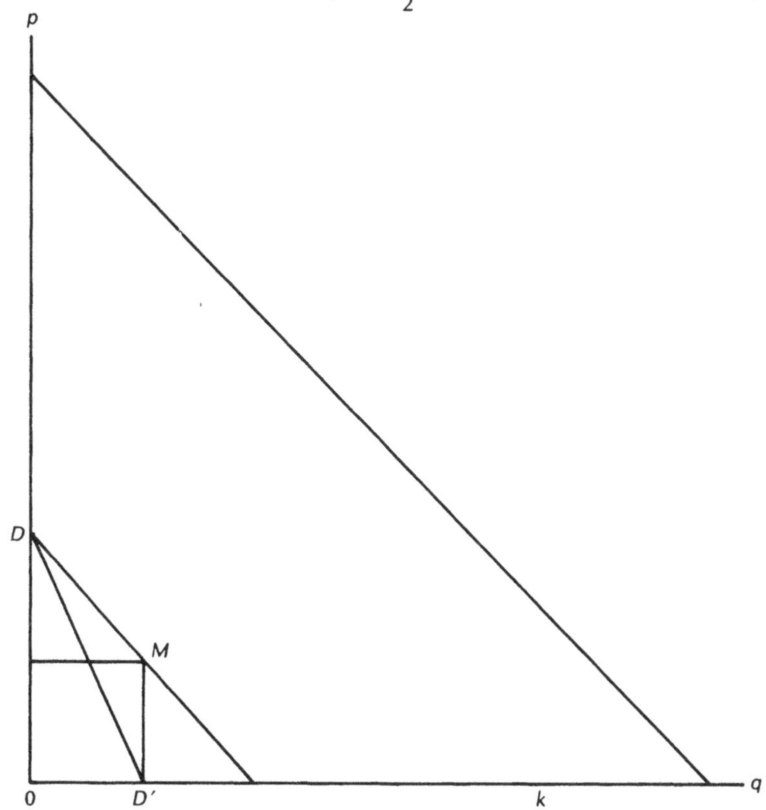

Figure 8.2. Duopoly with restricted capacity: case 2.

[Note that this example is illustrated by the point M when $k = 3a/4$ in Figure 8.2.]

Thus

$$\bar{p} = \frac{1}{k}\left(\frac{a-k}{2}\right)^2 \tag{8.5}$$

and

$$\bar{P} = \hat{P} = \left(\frac{a-k}{2}\right)^2 \tag{8.6}$$

We note if there were no excess or shortage of capacity (as would be the case in a competitive industry), $k_1 = k_2 = a/2$ and the range and profits would be

$$p = \frac{a}{8}, \quad \hat{p} = \frac{a}{4} \quad \text{and} \quad \bar{P} = \hat{P} = \frac{a^2}{16}. \tag{8.7}$$

We now come to a result that at first is surprising, then obvious. Suppose that we continue to shrink capacity in this market. When $k_1 = k_2 = a/3$, we see from Eqs. (8.4) and (8.5) that $\bar{p} = \hat{p} = a/3$; in other words, the range of fluctuations shrinks to a point and a pure strategy noncooperative equilibrium reappears. This occurs precisely at the value of the Cournot equilibrium point for the following reason: if the lower-priced firm is at capacity at $a/3$, this leaves $2a/3$ of the market for the other who will act as a monopolist and produce $a/3$.

In Figures 8.3 and 8.4 the Edgeworth range and its reason for disappearance at $k = a/3$ are illustrated. We first consider $k = a/2$. This case is shown in Figure 8.3. Each firm has just enough capacity to satisfy the whole market at monopoly price $p = a/2$. If the firms had a capacity of $k = a$, the curve $0PM$ would show the growth of revenues to the lower-priced firm as it raises price, always being able to satisfy total market demand. Since for prices $p < a/2$, a capacity of $k = a/2$ is insufficient to satisfy the market, the growth of revenue is given by the line $0M$ which intersects the curve $0PM$ at M.

The curve $0MMS$ shows the change in revenues as the higher priced firm increases its price. It has more than enough capacity to satisfy any demand that is left for it. Its revenues reach a maximum at the point MM and decline to zero if it continues to increase its price to $p = a/2$.

Consider the lower-priced firm charging $p = a/8$ and the higher priced firm charging a price a shade higher. The former will make a profit indicated by B and the latter will make a profit shown at D. Suppose that the firm with the higher price has an opportunity to change price. If he cuts his price to just below the price of the other firm, his profit will be approximately B. If he raises his price to $p = a/4$, his profit shown at MM will be as high as at B.

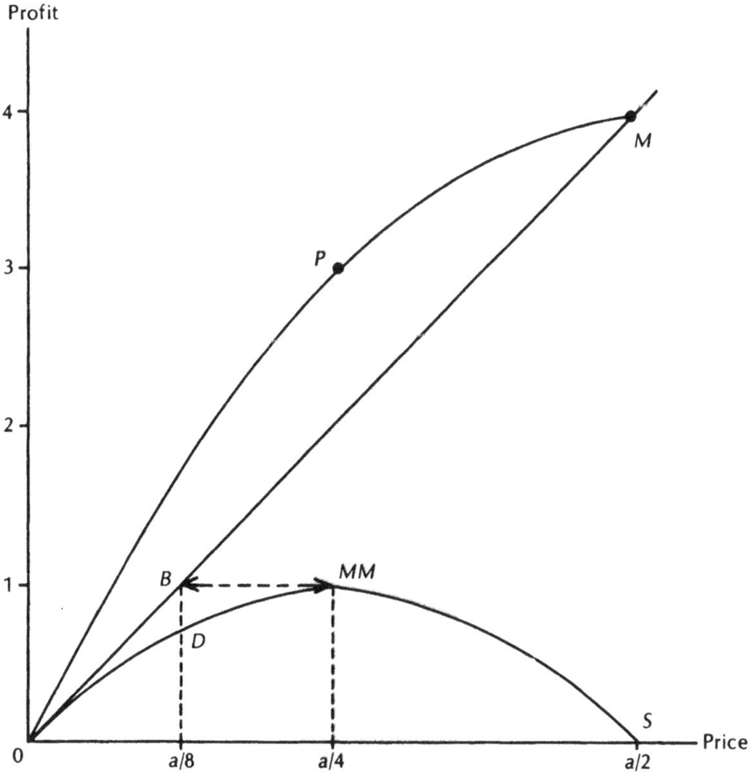

$k = a/2$

Figure 8.3. Duopoly profits with restricted capacity: case 1.

The Edgeworth range is given by *B* and *MM*. Furthermore if a firm pessimistically assumes that it will be undercut it should set its price at $p = a/4$. The point *MM* is also the $\max_{p_i} \min_{p_j} P_j$. It is the security level for either firm.[1]

Turning to Figure 8.4, the curve 0*PM* has the same meaning as in Figure 8.3. The line 0*MMC* is related to the line 0*M*, but here we observe that the individual firm no longer has enough capacity to satisfy the market at the monopoly price without capacity limits. At that price revenue is shown at *C*. We now note that this line goes through *MM*, but the Edgeworth cycle is determined by the horizontal distance from the point *MM* to the line 0*C*, which is now zero.

If one firm adopts a maxmin strategy the optimum reply for the other firm is to also adopt a maxmin strategy; hence they are in equilibrium.

1. In a more general model this is not the case. The relationships are more complicated. For further discussion and an example, see Shubik (1959b, ch. 5).

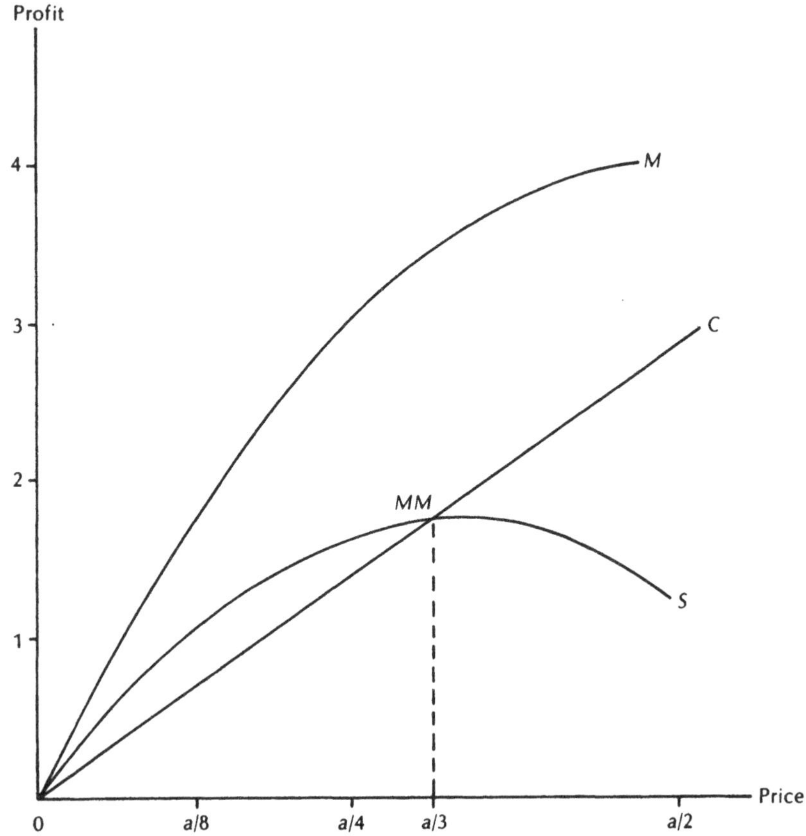

Figure 8.4. Duopoly profits with restricted capacity: case 2.

When $k = a/3$, the price and quantity noncooperative equilibria are the same. This holds true in the range $0 \leq k \leq a/3$, in which both firms will produce a capacity and price will be $\bar{p} = a - 2k$.

When the firms together have a total capacity less than a, the efficient point or competitive equilibrium[2] is no longer at $p = 0$ but becomes $p = a - 2k$. This price can be interpreted as the shadow price for the worth of an increment of new capacity.

Our results are summarized in Table 8.2. The entries in the table are prices for the appropriate solution. Instability in price competition is bounded from below by extreme excess capacity (if we regard $k_1 + k_2 = a$ as the correct amount of capacity then there is 100%

2. Without entry it is not quite correct to refer to the solution of Bertrand as the competitive equilibrium. It is better described as the efficient solution that assigns a shadow price to the value of capacity.

excess). It is also bounded above by a one-third shortage of capacity. Beyond this point capacity is so tight that both firms need not worry about undercutting.

Mixed Strategy Solution to the Price Game We now take the range $a/3 < k < a$ and investigate the nature of the mixed strategy in the price game. The mixed strategy could be interpreted as an indication of the distribution of prices in an unstable market. A mixed strategy equilibrium of a two-person game consists of a pair of probability distributions over the respective strategy spaces with the property that for each player any strategy chosen with positive probability must be optimal against the other player's probability mixture.

The proof of the shape of the mixed strategy is given elsewhere (Levitan and Shubik, 1972). The cumulative probability function is

$$\Phi_i(p) = \frac{kp - [\tfrac{1}{2}(a - k)]^2}{p(p + 2k - a)}. \tag{8.8}$$

The graph in Figure 8.5 shows distribution changes for the values $k = 0.9a$, $0.5a$, and $0.4a$.

What happens when capacities differ? If $k_1 \geq a$ and $k_2 \geq a$ or $k_1 \leq a/3$ and $k_2 \leq a/3$, a pure strategy exists. When the capacities are unequal but not in the ranges noted there is a mixed strategy solution. The mixed strategy is no longer continuous when the firms have unequal capacities but the firm with the larger capacity selects the upper point in its bidding range with a finite probability. The solution appears to be more of a mathematical curiosum than of economic interest.

We encountered the surprising result that as capacity shrunk the pure strategy equilibrium reappears at the Cournot equilibrium point. This is not general; it will be determined by the type of contingent demand structure that is postulated. For example, Beckmann (1965) using a contingent demand method originally suggested by Shubik (1959b), does not obtain a pure strategy equilibrium at the Cournot point.

The determination of the reappearance of the pure strategy equilibrium

Table 8.2 A comparison of solutions.

	Solution prices		
Capacity conditions	Cournot	Edgeworth-Bertrand	Efficient point
$k \geq a$	$a/3$	0	0
$a/2 < k < a$	$a/3$	fluctuation	0
$a/3 < k \leq a/2$	$a/3$	fluctuation	$a - 2k$
$0 < k \leq a/3$	$a - 2k$	$a - 2k$	$a - 2k$

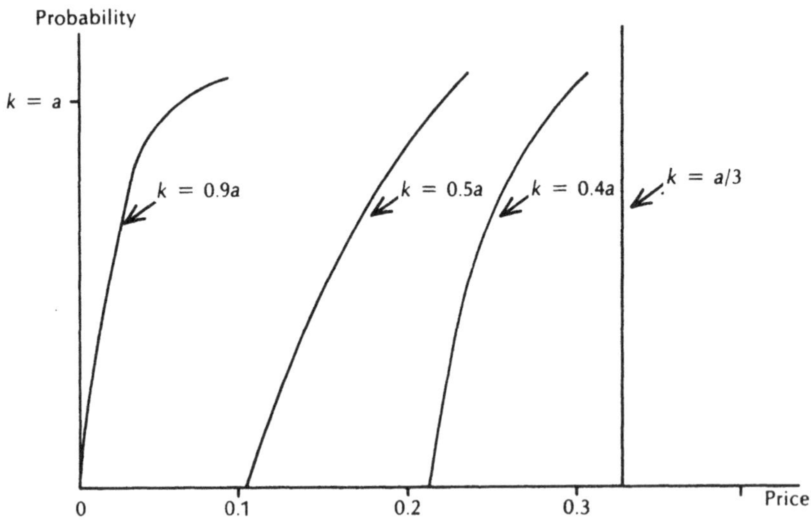

Figure 8.5. A mixed strategy solution.

depends upon the value of $\partial\Pi_i/\partial p_i$ at the point of potential equilibrium. The effect of moving capacity into the range $k_1 + k_2 \leq 2a/3$ puts a constraint on the derivative in the price-cutting direction. If a pure strategy equilibrium exists then both firms will be producing at capacity.

The test to see if capacity production is in equilibrium comes in the direction of raising price. In Figure 8.5 the contingent demand at the point *D* is *GD* which has an elasticity of 1 at *D*, hence there is no motivation to raise price.

Depending on the method used for the calculation of contingent demand the slope of the contingent demand for each firm may change as capacity is varied (when both are at capacity and the price is such that the market just clears). This point will be an equilibrium point if the elasticity of demand along the contingent demand curve is 1.

We have picked the most pessimistic method of calculating contingent demand. The high-priced portion of the demand curve is satisfied first. This gives a parallel translation of the demand inward as the shape of the contingent demand. In this case the Cournot point must be the point of demarcation for the reappearance of the pure strategy.

The actual shape of contingent demand cannot be specified generally from a priori reasoning. It will depend on priorities in service of customers and details of the summation of individual demand in specific markets. It is an important and complicated marketing problem which needs specific empirical investigation and model building.

The introduction of product differentiation does not appear to add any new qualitative results. The capacity limits on the conditions for the existence of a pure strategy equilibrium will change but beyond that the phenomena encountered will be qualitatively the same as it was without product differentiation.

Stockout Penalties

Introducing stockout penalties is a way to model adverse effects on future demand caused by the failure of a firm to supply product to some or all of its customers. This can at least capture some of the dynamic aspects of demand in a static model. An additional term must be introduced into the payoff functions to reflect this. An example of the effect of a stockout penalty on price instability and the Edgeworth cycle has been presented elsewhere (Shubik, 1968). It is shown that the instability may remain although the range of the cycle is changed.

Price-Quantity Duopoly with Inventory Carrying Costs

In Chapter 5 the price-quantity duopoly was described in which each firm commits to both a price and level of production simultaneously. If there are capacity constraints or if there are inventory carrying costs the pure strategy equilibrium will be destroyed.

Suppose that each firm, instread of choosing specific prices and production levels (p_i, q_i) were permitted to select probability distributions of the form

$$F_i(p, q) = \text{probability that} \quad p_i \leq p \quad \text{and} \quad q_i < q.$$

A mixed strategy equilibrium will exist if each firm can find a distribution such that if one knew the other's strategy neither would be motivated to change.

The demand faced by i given the strategy of j may be described as

$$d_i = \begin{cases} \alpha - \beta p_i & \text{for} \quad p_i < p_j, \quad c \leq p \leq \alpha/\beta, \\ (\alpha - \beta p_i)/\alpha & \qquad p_i = p_j, \\ \max(\alpha - q_j - \beta p_i, 0) & \qquad p_i > p_j. \end{cases} \tag{8.9}$$

If there were no costs attached to excess inventories both firms would produce to capacity. In the simplest case, any excess inventories are valueless. The payoff (shown for firm 1) can be expressed as

$$P_1 = \begin{cases} (p_1 - c)\min[q_1, \alpha - \beta p_1] \quad \text{for} \quad p_1 < p_2 \\ (p_1 - c)\max[0, \min(q_1, \alpha - q_2 - \beta p_1)] \\ \qquad - c\max[0, q_1 - (\alpha - q_2 - \beta p_1)] \quad \text{for} \quad p_1 > p_2. \end{cases} \tag{8.10}$$

For any choice of p_i there is associated a single q_j and the mixed strategy distribution need only be considered on price. The further details of modeling and proofs are given elsewhere (Levitan and Shubik, 1978). This distribution becomes

$$F_i(p, q) = \phi_i(p) = \frac{\beta p}{\beta p + c} \quad \text{and} \quad q = \alpha - \beta p$$

$$\text{for} \quad c \leq p \leq \alpha/\beta \quad \text{and} \quad c > 0. \qquad (8.11)$$

Although the mixed strategy equilibrium is somewhat unrealistic it is worth noting that even an attempt to define the price quantity duopoly model requires that inventory problems be considered. If the carrying of inventories is not free then even in a market with enough capacity for each to supply the whole market neither will wish to risk the losses.

We may observe immediately that for $c = 0$ the instability disappears unless there are capacity constraints. A brief calculation shows that the presence of potential inventory losses has the following feature: if the duopolists had no potential inventory losses and each had a capacity $k \geq \alpha - \beta c$ then there is a pure strategy equilibrium with $p_1 = p_2 = c$ and the firms make zero profits. The introduction of inventory loss possibilities destroys this equilibrium and leaves the firms with expected profits of zero but considerably lowers the welfare of the consumers.

The Noncooperative Price-Quantity Equilibrium with Random Demand

By making the reasonable assumption that production takes time we have shown in the previous section that a pure strategy equilibrium will not exist in a market with monopolistic competition. The Edgeworth model with price fluctuation appears to be the rule rather than the exception. In Figure 6.4, we demonstrated that even with product differentiation, capacity constraints could easily cause price instability. Here we stress that an inventory stockout can be reinterpreted as a short-term capacity constraint.

By adding the more realistic features of production capacity constraints and inventory considerations we destroyed equilibrium. Here it is shown that the addition of a further quite realistic complication helps to restore equilibrium. This complication is a degree of uncertainty in expected demand.

The general condition needed for stability is that the minimum inventories that the firms are induced to keep by consideration of demand fluctuations at a potential equilibrium point provide sufficient extra stock for each to make it unprofitable for either to change his price. The test for the condition is straightforward: solve the model for a pure strategy equilibrium by using the first-order conditions for a local maximum, then verify that no higher payoff can be achieved by having one firm change its price or quantity globally (i.e., over the whole range) given that the other firm is constrained by charging the price and supplying the quantity that is the candidate for the equilibrium.

In practice the stability will depend specifically upon the forms of the demand, production and inventory carrying cost functions. Here the analysis is limited to the duopoly model of Chapter 6. As the calculations are lengthy and have been given in detail elsewhere (Levitan and Shubik, 1971) we do not present them all here. Furthermore we use a somewhat simpler notation for this analysis than that found in Chapter 6:

Π = profit for the first (or distinguished) firm
p = price
q = demand
x = supply
ρ = inventory carrying cost per unit.

The same notation with an overbar (e.g., $\bar{p}$) represents the variables of the other firm.

By selecting the appropriate normalization for the symmetric game we can assume that production costs are zero (this involves replacing p by $p - c$ in the demand and profit functions). We may also set $\beta = 1$ without loss of generality. We use ϵ instead of α to indicate that the constant term has been replaced by a random variable.

Profits are given by

$$\Pi = p \min(q, x) - \rho \max(0, x - q). \tag{8.12}$$

A derivation of the curve *Degd'd''* shown in Figure 6.4 is given. We define the (normalized) ordinary linear demand curve to be

$$q(p, \bar{p}) = \epsilon - p - \gamma(p - \bar{p}), \tag{8.13}$$

which describes the demand for the distinguished firm when there are no stockouts or "price-outs." This function is represented by segment *gd'*. The segment *d'd''* describes the demand when the distinguished firm has priced its competitor out of the market. The demand on this segment is given by

$$\hat{\hat{q}}(p) = \frac{(1 + 2\gamma)(\epsilon - p)}{1 + \gamma}, \tag{8.14}$$

which is derived by assigning to the nondistinguished firm that price, $(\epsilon - \gamma p)/(1 + \gamma)$ which causes its demand to be exactly zero. Finally, the segment eg gives the demand of the distinguished firm when the nondistinguished firm's demand exceeds its supply, $\bar{x}$. The demand on eg is given by

$$q(p, \bar{x}) = \frac{(1 + 2\gamma)(\epsilon - p) - \gamma\bar{x}}{1 + \gamma}. \tag{8.15}$$

This function is derived similarly by substituting for $\bar{p}$ the price $(\epsilon - \bar{x} - \gamma p)/(1 + \gamma)$, which makes the second firm's demand equal to its supply.

It is easy to see that the demand function whose graph is $Degd'd''$ is equal to

$$q_T(p, \bar{p}, \bar{x}) = \max(0, \hat{q}(p, \bar{x}), \min[q(p, \bar{p}), \hat{\hat{q}}(p)])$$

and the actual sales of the distinguished firm are given by

$$s = \min(q_T, x) = \min(x, \min[0, \hat{q}, \min(q, \hat{\hat{q}})]).$$

We now introduce the random component into the demand by considering ϵ to be a random variable.

Expressing Eq. (8.12) as a function of actual sales, we can write the profit of the distinguished firm as

$$\Pi = ps - \rho(x - s) = (p + \rho)s - \rho x, \tag{8.16}$$

and by linearity, we can write the expectation of Π as

$$E(\Pi) = (p + \rho)E(s) - \rho x. \tag{8.17}$$

We may now derive expected sales. This involves deriving the conditions on ϵ for sales to have the values respectively of 0, $\hat{\hat{q}}$, q, $\hat{q}$, and x:

$$\begin{aligned} s = 0 &\Leftrightarrow \max[\hat{q}, \min(q, \hat{\hat{q}})] \leq 0 \\ &\Leftrightarrow \hat{q} \leq 0 \quad \text{and} \quad \min(q, \hat{\hat{q}}) \leq 0 \\ &\Leftrightarrow \hat{q} \leq 0 \quad \text{and} \quad (q \leq 0 \quad \text{or} \quad \hat{\hat{q}} \leq 0) \\ &\Leftrightarrow \epsilon < \max[\epsilon_2, \min(\epsilon_1, p)], \end{aligned}$$

where

$$\epsilon_1 = (1 + \gamma)p - \gamma\bar{p} \tag{8.18}$$

and

$$\epsilon_2 = p + \frac{\gamma}{1 + 2\gamma}\bar{x}. \tag{8.19}$$

In a similar manner we may calculate the other four conditions:

$$s = \hat{q} \Leftrightarrow \max(\epsilon_2, \epsilon_5) \leq \epsilon \leq \epsilon_7, \tag{8.20}$$

$$s = q \Leftrightarrow \max(\epsilon_1, \epsilon_4) \leq \epsilon \leq \min(\epsilon_5, \epsilon_6), \tag{8.21}$$

$$s = \hat{\hat{q}} \Leftrightarrow p \leq \epsilon \leq \min(\epsilon_3, \epsilon_4), \tag{8.22}$$

and

$$s = x \Leftrightarrow \epsilon \geq \max[\epsilon_3, \min(\epsilon_6, \epsilon_7)], \tag{8.23}$$

where

$$\epsilon_3 = p + \frac{1 + \gamma}{1 + 2\gamma}x, \tag{8.24}$$

$$\epsilon_4 = (1 + \gamma)\bar{p} - \gamma p, \tag{8.25}$$
$$\epsilon_5 = (1 + \gamma)\bar{p} - \gamma p + \bar{x}, \tag{8.26}$$
$$\epsilon_6 = (1 + \gamma)p - \gamma\bar{p} + x, \tag{8.27}$$

and

$$\epsilon_7 = p + \frac{\gamma\bar{x} + (1 + \gamma)x}{1 + 2\gamma}. \tag{8.28}$$

There always is a range for $s = 0$ and $s = x$ to be satisfied; the other cases however may be degenerate. The conditions for nondegeneracy must be specified. They are calculated in detail elsewhere (Levitan and Shubik, 1971).

Figure 8.6 illustrates the six regions in the decision space of the first player in which sales exhibit different qualitative characteristics. Expected profits have been denoted by $E(\Pi)$; for brevity Π will now be interpreted as $E(\Pi)$:

$$\Pi = (p + \rho)E(s) - \rho x, \tag{8.29}$$

where in, for example, R_1

$$E(s) = \int_p^{\epsilon_3} \frac{1 + 2\gamma}{1 + \gamma}(\epsilon - \gamma)dF + [1 - F(\epsilon_3)]x, \tag{8.30}$$

where F is the distribution function of the random variable ϵ.

The calculations required to derive the conditions for a pure strategy equilibrium are straightforward and tedious. The conditions for a symmetric pure strategy equilibrium imply that it must lie at the intersection of four of the six regions shown in Figure 8.6, where $p = \bar{p}$ and $x = \bar{x}$. At

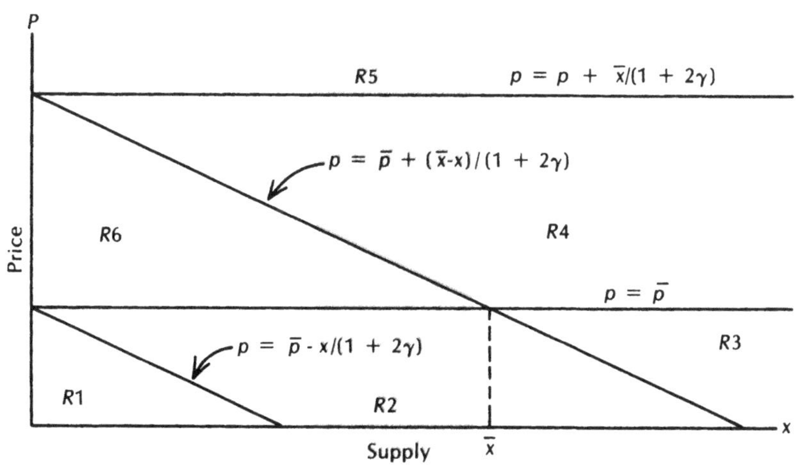

Figure 8.6. The decision space for the first duopolist.

this intersection we can show

$$\frac{\partial \Pi}{\partial p} = \int_p^{p+x} q dF + x[1 - F(p + x)]$$
$$- (1 + \gamma)(p + \rho)[F(p + x) - F(p)], \qquad (8.31)$$

$$\frac{\partial \Pi}{\partial x} = (p + \rho)[1 - F(p + x)] - \rho. \qquad (8.32)$$

In general we expect that the equations $\partial\Pi/\partial p = 0$ and $\partial\Pi/\partial x = 0$ can be solved to yield a unique pair $(\hat{p}, \hat{x})$. It remains to verify that $p = \bar{p} = \hat{p}$ and $x = \bar{x} = \hat{x}$ yield a pure strategy equilibrium, that is,

$$\Pi(\hat{p}, \hat{x}, \hat{p}, \hat{x}) \geq \Pi(p, x, \hat{p}, \hat{x}) \qquad \text{for all} \quad (p, x). \qquad (8.33)$$

This requires global conditions for a maximum.

We can illustrate the conditions for uniformly distributed demand: assume ϵ is distributed uniformly between α and $\alpha + \Delta$. Let $\bar{\epsilon} = \alpha + \Delta/2$. From Eqs. (8.31) and (8.32) together with first-order conditions we obtain

$$x = \frac{\Delta p}{p + \rho} + a - \hat{p} \qquad (8.34)$$

and

$$\frac{\partial \Pi}{\partial p} = \Phi(p)$$
$$= \int_p^{[\Delta p/(p+\rho)]+\alpha} (\epsilon - p) f(\epsilon) d\epsilon + \left(\frac{\Delta \rho}{p + \rho} + \alpha - p\right) \frac{\rho}{p + \rho}$$
$$- (\rho - p)(1 + \gamma) \left(\frac{p}{\rho + p} - F(p)\right) = 0, \qquad (8.35)$$

where

$$f(y) = 1/\Delta \qquad \text{if} \quad y \in [\alpha, \alpha + \Delta],$$
$$= 0 \qquad \text{otherwise.}$$

This equation has a unique solution $\hat{p}$ whose properties depend upon the value of $\Phi(\alpha) = -(1 + \gamma)\alpha$:

$$\text{If} \quad \Phi(\alpha) > 0, \quad \hat{p} = \frac{\alpha + \Delta + \rho + \sqrt{(\alpha + \Delta + \rho)^2 + 4(3 + 2\gamma)}}{2(3 + 2\gamma)} - \rho;$$

$$= 0, \quad \hat{p} = \alpha;$$
$$< 0, \quad \hat{p} = \text{unique positive root of}$$
$$2\rho^2\alpha + 2\rho[\Delta + 2\alpha - \rho(2 + \gamma)]p$$
$$+ [\Delta + 2\alpha - 4\rho(2 + \gamma)]p^2 - 2(2 + \gamma)p^3. \qquad (8.36)$$

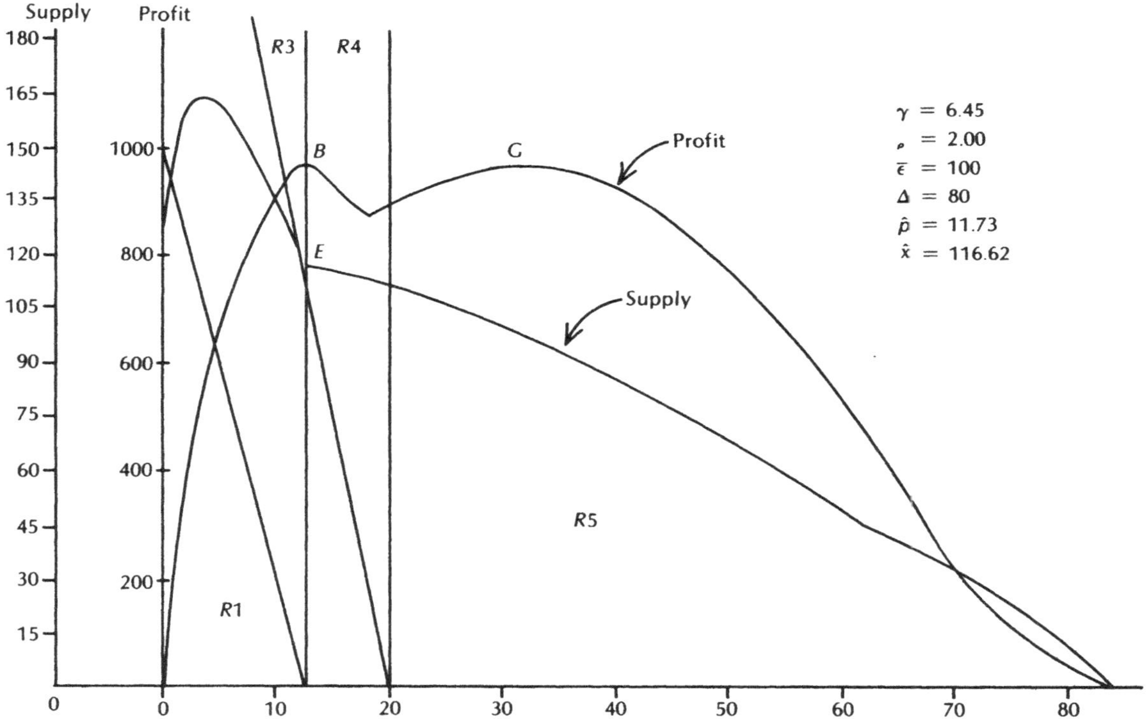

Figure 8.7. Profits, supply, and price.

Table 8.3 Maximal gamma for pure strategy equilibrium.

	$\Delta/\bar{\epsilon}$										
$\rho/\bar{\epsilon}$	0.025	0.05	0.10	0.20	0.30	0.40	0.50	0.60	0.70	0.80	0.90
0.0	0.71	1.02	1.55	2.54	3.61	4.84	6.30	7.50	8.31	8.83	9.15
0.010	0.70	1.01	1.51	2.43	3.38	4.40	5.53	6.47	7.02	7.32	7.44
0.020	0.69	0.99	1.47	2.33	3.19	4.08	5.01	5.81	6.25	6.45	6.50
0.030	0.69	0.97	1.44	2.25	3.04	3.83	4.63	5.34	5.71	5.86	5.86
0.040	0.68	0.96	1.41	2.18	2.91	3.62	4.33	4.97	5.31	5.42	5.39
0.050	0.67	0.94	1.38	2.12	2.80	3.45	4.09	4.67	4.98	5.07	5.03
0.060	0.66	0.93	1.35	2.06	2.70	3.30	3.88	4.43	4.72	4.80	4.74
0.070	0.65	0.92	1.33	2.01	2.61	3.18	3.71	4.21	4.50	4.56	4.50
0.080	0.65	0.90	1.30	1.96	2.53	3.06	3.56	4.03	4.31	4.37	4.29
0.090	0.64	0.89	1.28	1.91	2.46	2.96	3.43	3.86	4.14	4.20	4.12
0.100	0.63	0.88	1.26	1.87	2.40	2.88	3.31	3.72	4.00	4.05	3.97
0.110	0.63	0.87	1.24	1.84	2.34	2.79	3.21	3.59	3.86	3.92	3.84
0.120	0.62	0.86	1.23	1.80	2.29	2.72	3.11	3.47	3.75	3.80	3.72

0.130	0.61	0.85	1.21	1.77	2.24	2.65	3.03	3.37	3.64	3.70	3.61
0.140	0.61	0.84	1.19	1.74	2.19	2.59	2.95	3.27	3.54	3.60	3.52
0.150	0.60	0.83	1.18	1.71	2.15	2.54	2.88	3.19	3.45	3.52	3.43
0.200	0.58	0.79	1.11	1.59	1.97	2.30	2.59	2.85	3.07	3.18	3.11
0.300	0.54	0.73	1.01	1.42	1.73	2.00	2.23	2.42	2.59	2.73	2.74
0.400	0.51	0.68	0.94	1.30	1.57	1.80	2.00	2.16	2.30	2.41	2.50
0.500	0.48	0.65	0.88	1.21	1.46	1.66	1.84	1.98	2.10	2.20	2.28
0.600	0.46	0.62	0.83	1.14	1.37	1.56	1.72	1.85	1.96	2.05	2.13
0.700	0.44	0.59	0.80	1.08	1.30	1.47	1.62	1.74	1.85	1.94	2.01
0.800	0.43	0.57	0.76	1.03	1.24	1.40	1.54	1.66	1.76	1.84	1.91
0.900	0.41	0.55	0.74	0.99	1.19	1.34	1.48	1.59	1.68	1.76	1.83
1.000	0.40	0.53	0.71	0.96	1.14	1.29	1.42	1.53	1.62	1.70	1.76
1.100	0.39	0.52	0.69	0.93	1.10	1.25	1.37	1.48	1.57	1.64	1.71
1.200	0.38	0.50	0.67	0.90	1.07	1.21	1.33	1.43	1.52	1.59	1.66
1.300	0.37	0.49	0.65	0.87	1.04	1.17	1.29	1.39	1.47	1.55	1.61
1.400	0.36	0.48	0.64	0.85	1.01	1.14	1.25	1.35	1.43	1.51	1.57
1.500	0.36	0.47	0.62	0.83	0.99	1.11	1.22	1.32	1.40	1.47	1.53

When $\rho = 0$, that is, no inventory carrying costs exist, then if

$$\Phi(\alpha) \geq 0, \quad \hat{p} = \frac{\alpha + \Delta}{3 + 2\gamma}, \quad \hat{x} = \frac{2(1 + \gamma)}{3 + 2\gamma}(\alpha + \Delta);$$
$$< 0, \hat{p} = \frac{\bar{\epsilon}}{2 + \gamma}, \quad \hat{x} = \frac{1 + \gamma}{2 + \gamma}\bar{\epsilon} + \frac{\Delta}{2}. \tag{8.37}$$

For $\gamma \to \infty$, Eq. (8.37) becomes $\hat{p} = 0$, $\hat{x} = \alpha + \Delta$.

For deriving sufficient conditions, as analytic methods turned intractable, computation was used. Our method consisted of a numerical calculation, for any given set of values of the parameters $\bar{\epsilon}$, Δ, γ, and ρ, of the pair of values $\hat{p}$, $\hat{x}$ that satisfy the first-order conditions for equilibrium, and then a search of the function $\Pi(p)$ for the location of its maximum. If the maximum is at $\hat{p}$ then the solution is indeed an equilibrium. In Figure 8.7 we give sample curves, representing $\Pi(p)$ and $\hat{x}(p)$ in a case where $\Pi(p)$ has a secondary maximum exactly equal to $\Pi(p)$.

Noting that the model is homogeneous in the parameters $\bar{\epsilon}$, Δ, and ρ, we conducted a search for the maximal value of γ consistent with equilibrium over a range of the ratios $\rho/\bar{\epsilon}$ and $\Delta/\bar{\epsilon}$, and tabulated them in Table 8.3.[3] It was our computational experience that values of γ above the tabulated value always result in no equilibrium and those below always give an equilibrium solution in pure strategies.

This analysis of the stochastic duopoly model is concluded by examining the sensitivity of equilibrium price to changes in the values of the parameters. Since the result is quite apparent, we state without proof that $\bar{p}$ is a decreasing function of γ and an increasing function of $\bar{\epsilon}$. We must also deal with changes in the parameters ρ and Δ. The analysis differs depending on whether $\hat{p}$, the equilibrium price is greater than or less than or equal to α, the lower bound of the random variable ϵ.

For $\hat{p} \geq \alpha$, $\hat{p}$ increased with ρ for small ρ and then decreases for large ρ until the regime changes and $\hat{p} < \alpha$. When $\hat{p} > \alpha$ it is a decreasing function of both Δ and ρ.

Concluding Remarks

We have shown that if one improves the realism of the Chamberlinian model by adding inventory carrying costs and making the assumptions that production takes time that the pure strategy noncooperative equilibrium postulated by Chamberlin *never* exists. Instability of the type suggested by Edgeworth is all that remains.

By introducing a *further* complication into the model the equilibrium may be restored. This too is a step toward realism. It is the introduction of a random element to overall market size.

3. A more extensive tabulation is given in Levitan and Shubik (1971).

We leave as open problems the generalizations and the extension of our work to the n-person symmetric and nonsymmetric cases. Our experience with models of this variety indicates that it is a safe conjecture that our results go through for the nonsymmetric market model developed in Chapter 9. Although the results here are mathematically inelegant, our conjecture is of substantive and theoretical interest. While our proof of the *nonexistence* of the Chamberlinian equilibrium is perfectly general, our proof of the existence of equilibrium under uncertainty uses a specially simple example. To obtain a more general result, a more powerful and different type of mathematical approach is undoubtedly needed.

9 | A Nonsymmetric Oligopoly Model

THE ASSUMPTION OF SYMMETRY as presented in the duopoly and oligopoly models enables us to make vast simplifications in our analysis. Calculations that would otherwise be extremely difficult become simple with symmetry. Measures and indices of considerable use can be defined for the symmetric market; for example, there is a relatively straightforward answer to the question of how many is many, when the market is symmetric.

Unfortunately the actual world does not wish to fit our neater simpler models. Elegance and symmetry that might make the economics simpler and the mathematics neater seem to be the exception, not the rule, when studying market structures. We make no pretense at being able to supply a general analysis of the nonsymmetric market; however, at least it is possible to extend the model of Chapter 6 in a relatively natural way in such a manner that a minimal number of new parameters are introduced and the new model is relatively tractable. Several types of nonsymmetry are investigated. Each influences the steady-state solutions of a market model in considerably different ways.

The direct market interaction between firms is via price, advertising, or other product variation. To begin, we concentrate on price as the interactive variable. In terms of the model of Chapter 6 we replace the parameter $\beta\gamma$ by a more general array. Furthermore, it may be possible that the effectiveness of advertising or other product differentiation by one firm may differ from that of another. This might even reach to the levels of different consumer loyalty.

Internal to the individual firm, even excluding bureaucratic, behavioristic and other important differences that would have to be specified in the development of an adequate dynamic theory, are two important types of nonsymmetry that must be considered when comparing firms. Various fixed terms and initial conditions may differ. These can be overheads, capacity limitations, and initial monetary and physical assets. Variable costs may also differ; thus there may be different production costs, inventory carrying costs, costs of changing production or of borrowing new funds.

The model that we analyze can be described by the payoff or revenue functions as follows:

$$P_i = \left(p_i - c_i - \frac{k_i}{2}\right)(\beta w_i(V - p_i + \gamma(p_i - \bar{p})) - K_i, \qquad \text{for} \quad i = 1, 2, \ldots, n, \tag{9.1}$$

where

c_i = average unit cost of production for firm i,
k_i = unit inventory carrying costs per period,
K_i = overhead costs for firm i,
w_i = market share "weight" of firm i.

Equation (9.1) should be compared with (9.2) which shows the symmetric payoff:

$$P_i = \frac{1}{n}\left[p_i - c - \frac{k}{2}\right][\alpha - \beta p_i + \gamma(p_i - \bar{p})] - K, \qquad \text{for} \quad i = 1, 2, \ldots, n. \tag{9.2}$$

Since our first concern is steady-state solutions we can expect that any unit of inventory stays in inventory for one-half of the production period. In the steady-state solution in which production rates and inventory levels are in constant proportion we can simplify Eqs. (9.1) and (9.2) by defining a new cost:

$$c_i' = c_i + \frac{k_i}{2}.$$

(Note that in subsequent calculations, for simpler notation, we use c_i to represent the more complex c_i'.)

For convenience the notation in Eq. (9.1) differs somewhat from Chapter 6. Thus $V = \alpha/\beta$. This has a direct interpretation as the price at which demand is zero. The $\bar{p}$ is no longer the average price but $\bar{p} = \Sigma w_j p_j$ which is the weighted average price. The weights w_i are the means by which we introduce nonsymmetry into the demand structure as influenced by price. Nonsymmetry is introduced into the internal structure of the firms by c_i and K_i. It is assumed that $\Sigma w_i = 1$; hence, if all firms were to charge the same price, w_i would reflect the innate asymmetry of market share.

Disregarding considerations of product variation or advertising and concentrating only on price, we note that the cross-derivative of demand for firm i with respect to the price charged by firm j is given by

$$\frac{\partial d_i}{\partial p_j} = \beta \gamma w_i w_j.$$

This equation is symmetric. No income effect is assumed.

The cross-elasticity of demand for the product of firm i given a change in price of firm j is given by

$$E_{ij} = \frac{p_j \partial d_i}{d_i \partial p_j} = \frac{\gamma w_j p_j}{V - p_i + \gamma(p_i - \bar{p})}. \tag{9.3}$$

The own elasticity of the firm i is

$$E_{ii} = -\beta w_i(1 - \gamma + \gamma w_i)\left(\frac{p_i}{q_i}\right). \tag{9.4}$$

These results should be contrasted with the measures in Chapter 6, in which the overall market elasticity e is precisely the same.

The nonsymmetry somewhat lessens the value of E_{ij} and E_{ii} as measures in the sense that now there will be a different E_{ij} for every pair of competitors and a different E_{ii} for each firm.

Types of Nonsymmetry

Often it appears that the steady-state solutions are independent of initial conditions. From the viewpoint of mathematical analysis this observation is both true and trivial; however, when we consider the model as a reflection of some basic properties of an industry it is unreasonable to treat differences in initial conditions in so facile a manner. Although we do not propose to carry out a detailed analysis, we will indicate where the problems are and give a qualitative evaluation of the effect of different initial conditions.

Capacity Constraints If the capacity constraints are sufficiently large for all firms, a lack of symmetry will have no effect. This would imply that all firms have considerable excess capacity. If, however, some or all firms have relatively tight capacity constraints these will actually affect the solutions; cases of this have been studied in Chapter 5 and in detail elsewhere (Levitan and Shubik, 1971).

Given tight capacity constraints one might wish to consider two possibilities in such a market. Either new firms would tend to enter or the existing firms would buy more capacity. If it were possible for new firms to enter, the solutions would have to be modified to account for the increased number of competitors. If the firms were permitted to buy extra capacity then the final steady-state solutions would be the same; however, the transient state would be of considerable interest as it would reflect the appropriate investment policies to expand capacity.

Initial Financial and Real Assets In general, markets are not static. The availability of surplus funds usually permits a firm to explore new markets and vary its product or invest in plant. In the model under consideration there are neither new markets nor products and processes. The cash position does, however, play three roles. When capacities are low, cash flow considerations will influence the optimal investment policy. When cash position is high and capacities adequate, the transient dividend policy is influenced, although the steady-state is not. When cash positions are low, possibilities of bankruptcy appear and the steady-state solutions may be influenced by cut-throat competition (Shubik, 1959b; Shubik and Thompson, 1959; Telser, 1972). Different levels of real assets influence the transient policies in production scheduling and may indirectly cause all of the cases noted for different levels of liquid asset holdings.

General Overheads Differences in general overheads are reflected in Eq. (9.1) by K_i. This has generally no effect on transient or steady-state behavior except when overheads are so large that they cause negative cash flow, cut-throat competition, and exit problems.

Nonsymmetry in Variable Costs

In this model the variable costs are

1. Unit cost of production
2. The costs of change in production
3. Inventory carrying costs
4. The rate of interest.

Production and inventory costs directly modify the period payoff functions, as shown in Eq. (9.1). Differences in the costs of change in production do not change the steady-state solutions. The interest rate has two indirect influences in its relation to inventory carrying costs and further in relation to depreciation costs. If the firms are permitted to buy extra capacity, the parameters K_i are not quite constants as they will depend on capacity size and depreciation rates.

Efficiency and Optimality

Efficient and Competitive Solutions

The efficient production level for n firms each with possibly different average costs and capacities can be solved in a straightforward manner.

This is presented diagrammatically in Figure 9.1 for the simple case of identical products and the more general case is discussed. When products are identical the overall industry demand is given by *DF*. The marginal revenue to the industry as a whole with respect to price is given by *DE*. The marginal cost of production is shown by the step function curve starting at *B* and going to *C*. Efficient production takes place at *C* where marginal cost equals price.

For the differentiated market the result is essentially the same. Efficient production calls for each firm to set $p_i = c_i$. This can be obtained directly by considering the demand structure to be derived from an aggregate quadratic utility function then maximizing the consumer's welfare treating the production of the firms as technological constraints. This is equivalent to assuming that the firms are being run solely for the consumer's benefit.

A utility function that gives the demand structure shown in Eq. (9.1) is

$$U(q, m) = V\sum_{i=1}^{n} q_i - \frac{1}{2\beta(1+\gamma)}\left[\sum_{i=1}^{n} (q_i^2/w_i) + \gamma\left(\sum q_i\right)^2\right] + \lambda m. \quad (9.5)$$

where $q = (q_1, q_2, \ldots, q_n)$; m is the amount of money held by the consumers; λ is the marginal worth of money (which we will set equal to one).

Once more we must stress that we have used the phrase efficient production rather than competitive solution for while they may often appear to be the same mathematically the role of entry is not relevant to the former but is critical to the latter. In this chapter, entry is not discussed.

In the remaining sections of this chapter the mathematical derivation of

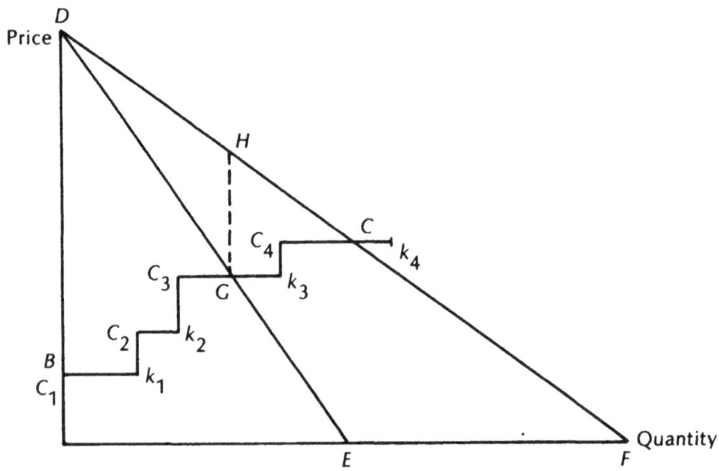

Figure 9.1. The efficient solution with many firms.

solutions is somewhat lengthy and tedious. For those with detailed interest in the construction of even the simplest of nonsymmetric market models we believe that this is necessary. Others may wish to merely examine the results and their application in the later sections of this chapter.

Joint Maximum Solution and Pareto Optimality

The profit of the ith firm is given by

$$\prod_i = \beta w_i(p_i - c_i)[V - p_i - \gamma(p_i - \bar{p})] \tag{9.6}$$

where

w_i = market share for the ith firm,
c_i = average cost of production for the ith firm,
$\bar{p}$ = weighted average of prices: $\left(\bar{p} = \sum_{j=1}^{n} w_j p_j,\ \sum w_j = 1\right)$.

Taking derivatives of Π_i with respect to i and j and setting them equal to zero we obtain

$$\frac{\partial \prod_i}{\partial p_i} = \beta w_i([p_i - c_i][-1 - \gamma(1 - w_i)] + [V - p_i - \gamma(p_i - \bar{p})])$$

$$= \beta w_i(V + [1 + \gamma - \gamma w_i]c_i - [2(1 + \gamma)p_i - \gamma w_i]p_i + \gamma\bar{p}) = 0; \tag{9.7}$$

$$\frac{\partial \prod_i}{\partial p_j} = \beta w_i(p_i - c_i)w_j\gamma = 0. \tag{9.8}$$

Summing Eq. (9.8) we obtain

$$\sum_{j \neq i} \frac{\partial \prod_j}{\partial p_i} = \gamma\beta w_i \sum w_j(p_j - c_j);$$

adding this to Eq. (9.7) gives

$$\frac{\partial \prod_j}{\partial p_i} = \beta w_i \Bigg[\gamma \sum_{j \neq i} w_j(p_j - c_j) + \gamma w_i(p_i - c_i) + V + (1 + \gamma)c_i$$

$$- 2(1 + \gamma)p_i - \gamma\bar{p}\Bigg]$$

$$= \beta w_i \left[\gamma \sum w_j(p_j - c_j) + V + (1 + \gamma)c_i - 2(1 + \gamma)p_i + \gamma\bar{p}\right]$$

$$= \beta w_i \left[V + (1 + \gamma)c_i - \sum w_j c_j - 2(1 + \gamma)p_i + 2\gamma\bar{p}\right] = 0. \tag{9.9}$$

Rearranging Eq. (9.9) and rewriting it in matrix notation gives

$$2[(1+\gamma)I - \gamma Sw]p = V\hat{1} + ((1+\gamma)I - \gamma Sw)c, \tag{9.10}$$

where

$$S = \begin{bmatrix} 1 & \cdots & 1 \\ \cdot & & \cdot \\ \cdot & & \cdot \\ \cdot & & \cdot \\ 1 & \cdots & 1 \end{bmatrix}, \quad \hat{1} = \begin{bmatrix} 1 \\ 1 \\ \cdot \\ \cdot \\ \cdot \\ 1 \end{bmatrix}, \quad w = \begin{bmatrix} w_1 & & & & \\ & w_2 & & & \\ & & \cdot & & \\ & & & \cdot & \\ & & & & \cdot \\ & & & & & w_n \end{bmatrix}, \quad p = \begin{bmatrix} p_1 \\ \cdot \\ \cdot \\ \cdot \\ p_n \end{bmatrix}, \quad c = \begin{bmatrix} c_1 \\ \cdot \\ \cdot \\ \cdot \\ c_n \end{bmatrix}.$$

This gives

$$\begin{aligned} p &= \frac{V}{2}[(1+\gamma)I - Sw]^{-1} + \frac{1}{2}c \\ &= \frac{V}{2}\frac{\gamma}{1+\gamma}\left(\frac{1}{\gamma}I + Sw\right)\hat{1} + \frac{1}{2}c \\ &= \frac{1}{2}c + \frac{V}{2}\frac{\gamma}{1+\gamma}(\hat{1} + \gamma\hat{1}) = \frac{1}{2}(c + V\hat{1}), \end{aligned} \tag{9.11}$$

or

$$p_i = \frac{c_i + V}{2}. \tag{9.12}$$

When the products are identical, this solution can be shown on a simple diagram. In Figure 9.1 the group of firms equate marginal cost and marginal revenue at the point *G* and charge the price at *H*.

No new conceptual difficulties appear in defining the Pareto optimal surface for the nonsymmetric market. As in Chapter 6, the efficient point is on the surface and favors the customer. The joint maximum is on the surface and favors the firms. In general, the noncooperative equilibrium outcomes in a nonsymmetric market are *not* Pareto optimal whereas they are in a symmetric market.

Noncooperative Equilibria

Price Equilibrium

Let the demand for the product of the *i*th firm be given by

$$d_i = w_i\beta[V - p_i - \gamma(p_i - \bar{p})]. \tag{9.13}$$

The revenue of the *i*th firm may be expressed as

$$\prod_i = (p_i - c_i)(w_i\beta[V - p_i - \gamma(p_i - \bar{p})]) - K_i. \tag{9.14}$$

For a noncooperative equilibrium we must solve the set of equations resulting from taking the derivatives of Eq. (9.14) with respect to p_i and setting them equal to zero:

$$\frac{\partial \prod_i}{\partial p_i} = \beta w_i(-[2(1+\gamma) - \gamma w_i]p_i$$

$$+ \gamma \sum w_j p_j + V + [1 + \gamma(1 - w_i)]c_i) = 0. \qquad (9.15)$$

All parameters and variables have been previously defined in this chapter. Using the matrix notation as in Eq. (9.10), we may express Eq. (9.15) in a more compact manner.

In equation (9.15) we divide by $\beta w_i \gamma$. For ease in notation, set

$$\Delta = \frac{2(1+\gamma)}{\gamma}.$$

Equation (9.15) can be written in matrix notation as

$$(\Delta I - W - SW)p = \frac{V}{\gamma}\,\hat{1} + \left(\frac{\Delta}{2} I - W\right) C, \qquad (9.16)$$

where

$$p = \begin{bmatrix} p_1 \\ \cdot \\ \cdot \\ \cdot \\ p_n \end{bmatrix}, \quad C = \begin{bmatrix} c_1 \\ \cdot \\ \cdot \\ \cdot \\ c_n \end{bmatrix}.$$

To solve Eq. (9.16) we evaluate the inverse of $(\Delta I - W - SW)$. The inverse is written as $[(\Delta W^{-1} - I) - S]W$. Let $\Delta W^{-1} - I = X$ and $[X - S]^{-1} = Y$. We may write δ_{ij} in terms of entries of X and Y as

$$\delta_{ij} = \sum_k (x_i \delta_{ik} - 1) y_{kj}$$

$$= x_i y_{ij} - \sum_k y_{kj}. \qquad (9.17)$$

Call $R_j = \Sigma_k y_{kj}$. Since Y is symmetric we write

$$y_{ij} = \frac{R_j + \delta_{ij}}{x_i} = \frac{R_i + \delta_{ji}}{x_j}. \qquad (9.18)$$

If $i \neq j$, then $R_i/x_j = R_j/x_i$; hence

$$R_i = \frac{R_j x_i}{x_j}. \qquad (9.19)$$

But $R_j x_j$ is a constant, which we may call Q; we rewrite Eq. (9.19) as

$$R_i = \frac{Q}{x_i}. \tag{9.20}$$

Returning to Eq. (9.18) we have

$$y_{ij} = \frac{\dfrac{Q}{x_j}}{x_i} = \begin{cases} \dfrac{Q}{x_i x_j} & \text{for} \quad i \neq j, \\ \\ \dfrac{Q}{x_1^2} + \dfrac{1}{x_i} & \text{for} \quad i = j. \end{cases} \tag{9.21}$$

Call

$$Z = \begin{bmatrix} \dfrac{1}{x_1} & & 0 \\ & \ddots & \\ 0 & & \dfrac{1}{x_n} \end{bmatrix}, \quad z_i = \frac{1}{x_i}.$$

We can write

$$[X - S]^{-1} = QZSZ + Z. \tag{9.22}$$

Evaluating Q:

$$\frac{Q}{x_i} = R_i = \sum_j y_{ij} = \sum_j \frac{Q}{x_i x_j} + \frac{1}{x_i};$$

hence

$$Q = Q \sum_j \frac{1}{x_j} + 1 = \frac{1}{1 - \sum_j (1/x_j)} = \frac{1}{1 - \sum_j z_j}. \tag{9.23}$$

Now we have the inverse of $X - S$; hence the prices can be written as

$$p = W^{-1}(QZSZ + Z)\left[\frac{V}{\gamma}\hat{1} + \left(\frac{\Delta}{2}I - W\right)c\right]. \tag{9.24}$$

Remembering that $z_i = w_i/(\Delta - w_i)$ and $\Sigma z_j = 1 - 1/Q$ we may write a specific price p_i as

$$p_i = \frac{z_i}{w_i}\left(\frac{V}{\gamma}(Q\Sigma z_j + 1) + \frac{1}{2}[(\Delta - 2w_i)c_i + Q\Sigma z_j(\Delta - 2w_j)c_j]\right), \tag{9.25}$$

which simplifies to

$$p_i = \frac{1}{\Delta - w_i}\left(\frac{V}{\gamma}Q + \frac{1}{2}[(\Delta - 2w_i)c_i + Q\Sigma z_j(\Delta - 2w_j)c_j]\right). \quad (9.26)$$

There are two special cases and a check that should be made. These are when (a) the costs are the same, that is, $c_i = c$ and (b) market conditions are the same, that is, $w_i = w$. We assume these values and investigate the resultant specification to see if it gives us the formula in the symmetric case.

A Special Case: Equal Costs If we assume that $c_i = c$ but that the w_i are different, Eq. (9.26) may be written as

$$p_i = \frac{1}{\Delta - w_i}\left(\frac{V}{\gamma}Q + \frac{c}{2}[(\Delta - 2w_i) + Q\Sigma z_j(\Delta - 2w_j)]\right). \quad (9.27)$$

We wish to sum the term $\Sigma z_j(\Delta - 2w_j)$. This can be written as

$$\Sigma z_j(\Delta - w_j) - \Sigma z_j w_j, \quad (9.28)$$

or as

$$\Delta\Sigma z_j = 2\Sigma z_j w_j. \quad (9.29)$$

However, by setting $z_j = w_j/(\Delta - w_j)$ in the first part of Eq. (9.28) we have immediately that $\Sigma z_j(\Delta - w_j) = 1$. Furthermore, as $Q = 1/(1 - \Sigma z_j)$, then $\Sigma z_j = (Q - 1)/Q$, which may be substituted in the first part of Eq. (9.29). Setting Eq. (9.28) equal to (9.29), we have

$$1 - \Sigma z_j w_j = \Delta\left(\frac{Q-1}{Q}\right) - 2\Sigma z_j w_j, \quad (9.30)$$

or

$$\Sigma z_j w_j = \Delta\left(\frac{Q-1}{Q}\right) - 1; \quad (9.31)$$

hence

$$\Sigma z_j(\Delta - 2w_j) = 2 - \Delta\left(\frac{Q-1}{Q}\right). \quad (9.32)$$

Substituting Eq. (9.32) into (9.27), we obtain

$$\begin{aligned} p_i &= \frac{1}{\Delta - w_i}\left(\frac{VQ}{\gamma} + \frac{c}{2}[\Delta - 2w_i + 2Q - \Delta(Q - 1)]\right) \\ &= \frac{1}{\Delta - w_i}\left(\frac{V}{\gamma}Q + \frac{c}{2}[2\Delta - 2w_i + Q(2 - \Delta)]\right). \end{aligned} \quad (9.33)$$

This equation can be rewritten in a better form as

$$\boxed{p_i = c + \frac{1}{\Delta - w_i}\left(\frac{V - c}{\gamma}\right).} \tag{9.34}$$

A Special Case: A Symmetric Market but Unequal Costs Let $w_i = w = 1/n$, then $z_i = (1/n)/(\Delta - 1/n)$ and $Q = (n\Delta - 1)/(n\Delta - 1 - n)$. We may write Eq. (9.26) as

$$\begin{aligned} p_i &= \frac{n}{n\Delta - 1}\left(\frac{V}{\gamma} Q + \frac{1}{2}\left[\left(\frac{n\Delta - 2}{n}\right) c_i + Q\left(\frac{1}{n\Delta - 1}\right)\left(\frac{n\Delta - 2}{n}\right) c_j\right]\right) \\ &= \frac{nV}{(n\Delta - 1 - n)\gamma} + \left(\frac{n}{n\Delta - 1}\right)\left(\frac{n - 2}{n}\right) \\ &\quad \left[c_i + \frac{1}{n\Delta - 1 - n}\Sigma c_j\right]. \end{aligned} \tag{9.35}$$

Substituting $\Delta = 2(1 + \gamma)/\gamma$, we obtain

$$\boxed{p_i = \frac{V}{2 + \left(\frac{n - 1}{n}\right)\gamma} + \frac{1 + \gamma\left(\frac{n - 1}{n}\right)}{2 + \left(\frac{2n - 1}{n}\right)\gamma}\left[c_i + \frac{\gamma}{2 + \frac{n - 1}{n}\gamma}\bar{c}.\right]} \tag{9.36}$$

The Symmetric Case: Two Checks In Eq. (9.34) we use the added condition of $w_i = w = 1/n$. This gives $z_i = 1/(n\Delta - 1)$ hence $Q = (n\Delta - 1)/(n\Delta - 1 - n)$. We obtain

$$\begin{aligned} p_i &= c + \left(\frac{n\Delta - 1}{n\Delta - 1 - n}\right)\left(\frac{n}{n\Delta - 1}\right)\left(\frac{V - c}{\gamma}\right) \\ &= c + \left(\frac{1}{2[(1 + \gamma)/\gamma] - (1/n) - 1}\right)\left(\frac{V - c}{\gamma}\right). \end{aligned} \tag{9.37}$$

This simplifies to

$$\boxed{p_i = \frac{V + c\left(1 + \frac{n - 1}{n}\gamma\right)}{2 + \frac{n - 1}{n}\gamma}.} \tag{9.38}$$

This is the symmetric solution as shown in Chapter 6. In Eq. (9.36) we use the added condition that $c_i = c$. This gives

$$p_i = \frac{V}{2 + \left(\frac{n - 1}{n}\right)\gamma} + \left[\frac{1 + \gamma\left(\frac{n - 1}{n}\right)}{2 + \left(\frac{2n - 1}{n}\right)\gamma}\right]\left[\frac{c\left(2 + \frac{n - 1}{n}\gamma + \gamma\right)}{2 + \frac{n - 1}{n}\gamma}\right]$$

$$= \frac{V + c\left(1 + \frac{n-1}{n}\right)\gamma}{2 + \frac{n-1}{n}\gamma}. \tag{9.39}$$

PRODUCTION OR QUANTITY EQUILIBRIUM

The nonsymmetric noncooperative extension of the Cournot model can be calculated by expressing the payoff with quantity rather than price as the independent variable.

As before, q is the vector $(q_1, q_2, \ldots, q_n)$ of quantities offered by the firm.

$$q = \beta W(V\hat{1} - [(1+\gamma)I - \gamma SW]p), \tag{9.40}$$

hence

$$\begin{aligned} p &= ((1+\gamma)I - \gamma SW)^{-1}\left[V\hat{1} - \frac{1}{\beta}W^{-1}q\right] \\ &= \frac{1}{\gamma}W^{-1}\left(\frac{1+\gamma}{\gamma}W^{-1} - S\right)^{-1}\left(V\hat{1} - \frac{1}{\beta}W^{-1}q\right) \\ &= \frac{1}{1+\gamma}(I + SW)\left(V\hat{1} - \frac{1}{\beta}W^{-1}q\right) \end{aligned}$$

or

$$p = V\hat{1} - \frac{1}{\beta(1+\gamma)}(W^{-1} + \gamma S)q. \tag{9.41}$$

The ith payoff may be expressed as

$$\begin{aligned} \prod_i &= q_i(p_i - c_i) \\ &= q_i\left(V - \frac{1}{\beta(1+\gamma)}\left(\frac{q_i}{w_i} + \gamma\,\Sigma q_j\right) - c_i\right). \end{aligned} \tag{9.42}$$

Differentiating with respect to quantity we obtain

$$\begin{aligned} \frac{\partial \sum_i}{\partial q_i} &= V - \left(\frac{1}{\beta(1+\gamma)}\right)\left(\frac{q_i}{w_i} + \gamma\Sigma q_j\right) - c_i \\ &\quad - \left(\frac{q_i}{\beta(1+\gamma)}\right)\left(\frac{1}{w_i} + \gamma\right). \end{aligned} \tag{9.43}$$

Setting Eq. (9.43) equal to zero we have

$$V - c_i - \left(\frac{1}{\beta(1+\gamma)}\right)\left(\left[\frac{2}{w_i} + \gamma\right]q_i + \gamma\Sigma q_j\right) = 0. \tag{9.44}$$

Writing this in matrix notation yields

$$\frac{\gamma}{\beta(1+\gamma)}\left(\frac{1}{\gamma}(2W^{-1}+\gamma I)+S\right)q = V\hat{\imath} - c. \tag{9.45}$$

We note that

$$(Z^{-1}+S)^{-1} = (Z - XZSZ), \tag{9.46}$$

where

$$X = \frac{1}{1+\Sigma z_i}.$$

In this case [Eq. (9.45)]

$$z_i = \frac{\gamma w_i}{2+\gamma w_i}.$$

Specifically,

$$X = \frac{1}{1+\gamma\Sigma w_j/(2+\gamma w_j)}.$$

Solving for q, we obtain

$$q = \frac{\beta(1+\gamma)}{\gamma}(Z - XZSZ)(V\hat{\imath} - c) \tag{9.47}$$

and

$$p = V\hat{\imath} - \frac{1}{\gamma}(W^{-1}+\gamma S)(Z - XZSZ)(V\hat{\imath} - c). \tag{9.48}$$

Multiplying out the factors on the right-hand side of Eq. (9.48) yields

$$p = V\hat{\imath} - \frac{1}{\gamma}(W^{-1}+\gamma S - XW^{-1}ZS - \gamma X\hat{\imath}\hat{\imath}^T Z\hat{\imath}\hat{\imath}^T)(Z)(V\hat{\imath} - c); \tag{9.49}$$

however,

$$\hat{\imath}^T Z\hat{\imath} = \frac{1-X}{X},$$

hence

$$p = V\hat{\imath} - \frac{1}{\gamma}[W^{-1}+\gamma S - XW^{-1}ZS - \gamma(1-X)S](Z)(V\hat{\imath} - c) \tag{9.50}$$

or

$$p = V\hat{\imath} - \frac{1}{\gamma}[W^{-1}+X(\gamma I - W^{-1}Z)S](Z)(V\hat{\imath} - c), \tag{9.51}$$

which simplifies to

$$p = \left(\hat{\imath} - \frac{1}{\gamma}[W^{-1}Z\hat{\imath} + X(\gamma I - W^{-1}Z)SZ\hat{\imath}]\right) + \frac{1}{\gamma}[W^{-1} + X(\gamma I - W^{-1}Z)S]ZC. \tag{9.52}$$

Writing $S = \hat{\imath}\hat{\imath}^T$ in Eq. (9.52) gives us $SZ\hat{\imath} = \hat{\imath}\hat{\imath}^T Z\hat{\imath} = \hat{\imath}(1 - X/X)$; hence

$$p = V\left(\hat{\imath}[1 - (1 - X)] - \frac{1}{\gamma}W^{-1}Z\hat{\imath}[\hat{\imath} - (1 - X)]\right) + \frac{1}{\gamma}W^{-1}ZC + \frac{X}{\gamma}(\gamma I - W^{-1}Z)SZC \tag{9.53}$$

or

$$p = XV\left(1 - \frac{1}{\gamma}Z1\right) + \left(\frac{c_i}{2 + \gamma w_i}\right) + X(\gamma I - W^{-1}Z)\Sigma\frac{w_j c_j}{2 + \gamma w_j}; \tag{9.54}$$

therefore

$$p_i = XV\left(\frac{1 + \gamma w_i}{2 + \gamma w_i}\right) + \left(\frac{c_i}{2 + \gamma w_i}\right) + X\gamma\left(\frac{1 + \gamma w_i}{2 + \gamma w_i}\right)\Sigma\frac{w_j c_j}{2 + \gamma w_j} \tag{9.55}$$

or

$$\boxed{p_i = X\left(V + \gamma\Sigma\frac{w_j c_j}{2 + \gamma w_j}\right)\left(\frac{1 + \gamma w_i}{2 + \gamma w_i}\right) + \left(\frac{c_i}{2 + \gamma w_i}\right).} \tag{9.56}$$

We may check some special cases.

A. Suppose $\gamma = 0$; this implies $X = 1$:

$$\begin{aligned} p &= \tfrac{1}{2}V\hat{\imath} + \tfrac{1}{2}c \\ &= \tfrac{1}{2}(V\hat{\imath} + c), \end{aligned}$$

which is the monopoly solution.

B. Suppose $\gamma \to \infty$; this implies $z_1 \to 1$, hence $X \to 1/n + 1$:

$$\begin{aligned} p &\to \frac{1}{n + 1}(V + \Sigma c_j) + 0 \\ &= \frac{V}{n + 1} + \frac{n}{n + 1}\bar{c}, \end{aligned}$$

where $\bar{c}$ is the average cost.

C. Suppose $w_i = 1/n$; this formula becomes

$$p_i = \left(\frac{n+\gamma}{2n+(n+1)\gamma}\right)\left(V + \frac{\gamma n}{2n+\gamma}\bar{c}\right) + \left(\frac{nc_i}{2n+\gamma}\right). \tag{9.57}$$

We now compare the noncooperative equilibrium obtained from regarding price as the independent variable or quantity as the independent variable. For $n = 1$ we see below from the two general symmetric market formulas that we obtain the same result. Similarly for $n \to \infty$ we obtain the same limits (when all $c_i = \bar{c}$).

$$p_i(\text{Edgeworth}) = \left[\frac{n}{2n+(n-1)\gamma}\right]\left[V + \left(\frac{n+(n-1)\gamma}{2n+(2n-1)\gamma}\right)\gamma\bar{c}\right] + \left[\frac{(n+(n-1)\gamma)\bar{c}}{2n+(2n-1)\gamma}\right]; \tag{9.58}$$

$$p_i(\text{Cournot}) = \left[\frac{n+\gamma}{2n+(n+1)}\right]\left[V + \frac{\gamma n}{2n+\gamma}\bar{c}\right] + \left[\frac{n}{2n+\gamma}\bar{c}\right]. \tag{9.59}$$

For $n = 1$

$$p_i(\text{Edgeworth}) = \frac{1}{2}\left(V + \frac{\gamma}{2+\gamma}\bar{c}\right) + \frac{\bar{c}}{2+\gamma} = \frac{1}{2}(V+\bar{c});$$

$$p_i(\text{Cournot}) = \frac{1+\gamma}{2(1+\gamma)}\left(V + \frac{\gamma}{2+\gamma}\bar{c}\right) + \frac{\bar{c}}{2+\gamma} = \frac{1}{2}(V+\bar{c}).$$

For $n \to \infty$

$$p_i(\text{Edgeworth}) = \frac{1}{2+\gamma}\left[V + \frac{1}{2}\left(\frac{1+\gamma}{1+\gamma}\right)\gamma\bar{c}\right] + \frac{\bar{c}}{2} = \frac{V}{2+\gamma} + \bar{c}\left[\frac{\gamma+1}{\gamma+2}\right];$$

$$p_i(\text{Cournot}) = \frac{1}{2+\gamma}\left(V + \frac{\gamma}{2}\bar{c}\right) + \frac{\bar{c}}{2} = \frac{V}{2+\gamma} + \bar{c}\left(\frac{\gamma+1}{\gamma+2}\right).$$

We note that for $\gamma \to \infty$ these equations both give $p_i = \bar{c}$.

Returning to Eq. (9.58) and (9.59) for $c_i = c$, we may observe that for $n > 1$ and $\gamma > 0$

$$p(\text{Cournot}) > p(\text{Edgeworth}).$$

This follows from observing that

$$\frac{n+\gamma}{2n+(n+1)\gamma} > \frac{n}{2n+(n-1)\gamma}$$

and that price is a weighted average of costs c_i and V; however $V \geq \bar{c}$, and V appears with a larger weighting in Eq. (9.59).

Other Solutions

Generalized Beat-the-Average Solution

In Chapter 6 the beat-the-average solution was noted as the symmetric n-firm extension of the highly competitive maxmin-the-difference solution for a two-firm market. When symmetry is not present, unbridled beat-the-average behavior does not appear to be reasonable, especially when some firms may be inherently at such a disadvantage that this type of behavior would not only be hostile but suicidal.

A candidate for an extremely competitive solution to be applied to an n-firm nonsymmetric market is that which we call "maximize the share of industry profits," or

$$\max \frac{P_i}{\sum_{j=1}^{n} P_j} \quad \text{or} \quad \frac{(p_i - c_i)q_i}{\sum_{j=1}^{n} (p_j - c_j)q_j}. \tag{9.60}$$

When all firms are symmetric we may observe immediately that this solution will yield the same result as beat-the-average. In this sense it can be regarded as a generalization of beat-the-average. However there are some important conceptual differences that must not be overlooked. The dimension of the beat-the-average solution is money. The "maximize the share of industry profits" solution has dimensions of money divided by money, that is, it is dimension-free or is a pure number.

If we contemplate a market in which the firms are completely symmetric except for differences in overheads, the beat-the-average solution will be unchanged, since in the short run the overhead costs are not strategically significant. The profit-share solution however would be affected by the change in overheads. This appears to be unreasonable, hence when using this solution it is desirable to make the modification that fixed costs or revenues are excluded. The method for calculating this solution is somewhat messy, but is outlined in a following section of this chapter where it is shown to be connected with the calculation of another solution.

Maximize-Market-Share Solution

The maximize-market-share solution has already been noted in Chapter 6. However it is more natural to consider this solution when firms are not symmetric and to contrast it with the maximization of profit share. A glance at Eq. (9.60) shows that when $c_i = 0$ the two solutions become the same and amount to

$$\max \frac{R_i}{\sum_{j=1}^{n} R_j} \quad \text{or} \quad \frac{p_i q_i}{\sum_{j=1}^{n} p_j q_j}. \tag{9.61}$$

This solution can be calculated easily by calculating the profit share solution with $c_i = 0$ then subtracting $c_i q_i$ from the revenues of the ith firm. Some examples are provided in a later section.

We note that prices and outputs are not influenced by any changes in costs. From the viewpoint of the firm as a coordinated whole this is undesirable. However a defense for considering this solution can be offered on behavioral or organizational grounds. Usually (but not always) the marketing department's efforts to increase market share are correlated with the welfare of the firm as a whole. When costs are perceived to be too high a special (and separate) effort is made by the manufacturing department to cut them.

Internal control, communications, and incentive problems are often excluded from economic models of the firm. When they are included, actions that might be regarded as irrational from a simplified economic viewpoint may be consistent with rational behavior that takes into account these added organizational factors.

Measures of Structure and Behavior

Efficiency and Consumer Surplus

In a nonsymmetric market, consumer surplus becomes an even more valuable measure than it is with symmetry present. Several measures can be suggested for measuring the efficiency of market performance with symmetry. Without symmetry the measurement becomes more difficult but consumer surplus remains a useful index.

In Eq. (9.5) a quadratic utility function yielding market demand structure was presented with a linear money component included directly in the utility function. If we wish we could conceptualize this somewhat differently excluding money from the utility function and solving a constrained maximization problem. Let

$$U(q) = \tfrac{1}{2} q^t A q + b^t q, \tag{9.62}$$

where q is the vector of quantities of commodities and q^t is the transpose of q. Given p, a vector of prices for all commodities, then

$$U'(q) = Aq + b = \lambda p, \tag{9.63}$$

where λ is a Lagrangian multiplier. From Eq. (9.63) and the demand factor in Eq. (9.1), we obtain

$$\begin{aligned} q &= A^{-1}(\lambda p - b) \\ &= \beta w[V - p - \gamma(p - Swp)] \\ &= \beta w[(\gamma S w - (1 + \gamma) I) p + V\hat{1}], \end{aligned}$$

yielding

$$\lambda A^{-1} = \beta w(\gamma S w - (1 + \gamma) I)$$

and

$$A^{-1}b = \beta V w I;$$

hence

$$A = \frac{\lambda}{\beta}(\gamma S w - (1 + \gamma) I)^{-1} w^{-1}$$
$$= \frac{\lambda}{\gamma\beta} w^{-1}\left(S - \frac{1 + \gamma}{\gamma} w^{-1}\right)^{-1} w^{-1} \qquad (9.64)$$

and

$$b = -\lambda V(\gamma S w - (1 + \gamma) I)^{-1}\hat{\imath}. \qquad (9.65)$$

Call $[(1 + \gamma)/\gamma]w^{-1} = Z$, then from Eq. (9.64),

$$A = \frac{\lambda}{\gamma\beta} w^{-1}(S - Z)^{-1} w^{-1}. \qquad (9.66)$$

It can be shown that

$$(S - Z)^{-1} = -\left(\frac{1}{1 + \gamma} w + \frac{\gamma^2}{1 + \gamma} wSw\right),$$

yielding

$$A = -\frac{\lambda}{\beta}\left(\frac{1}{1 + \gamma} w^{-1} + \frac{\gamma}{1 + \gamma} S\right) = -\frac{\lambda}{(1 + \gamma)\beta}(w^{-1} + \gamma S), \qquad (9.68)$$

and

$$b = -\beta V A w \hat{\imath} = \frac{\lambda V}{1 + \gamma}(w^{-1} + \gamma S) w \hat{\imath} \qquad (9.69)$$
$$= \frac{\lambda V}{(1 + \gamma)}[(1 + \gamma)\hat{\imath}] = \lambda V \hat{\imath},$$

yielding

$$\frac{U(q)}{\lambda} = V \sum_{i=1}^{n} q_i - \frac{1}{2\beta(1 + \gamma)}\left[\sum_{i=1}^{n}\left(\frac{q_i^2}{w_i}\right) + \gamma \sum_{i=1}^{n} q_i\right]^2. \qquad (9.70)$$

This may be compared with Eq. (9.5).

For some purposes it may be easier to calculate consumer surplus from price information directly; hence it is handy to have a utility function measured in terms of price. From Eqs. (9.68) and (9.69) we have

$$U(q) = -\frac{\lambda}{2\beta} q^t \left(\frac{1}{1 + \gamma} w^{-1} + \frac{\gamma}{1 + \gamma} S\right) q + \lambda V \hat{\imath}^t q. \qquad (9.71)$$

We may write the demand function as

$$q = r - Qp, \qquad (9.72)$$

where $r = \beta V w \hat{1}$ and $Q = \beta w[(1+\gamma)I - \gamma S w]$, as can be seen from differentiating Eq. (9.13) and obtaining

$$b - Aq = \lambda p \quad \text{or} \quad q = A^{-1}(b - \lambda p) = r - Qp.$$

Let the utility function in terms of price be $\phi(p)$; then

$$\phi(p) = b^t q - \tfrac{1}{2} q^t A q. \tag{9.73}$$

Divide by λ and substitute for q to obtain

$$\begin{aligned}\frac{\phi(p)}{\lambda} &= r^t Q^{-1}(r - Qp) - \frac{1}{2}(r - Qp)^t Q^{-1}(r - Qp)\\ &= r^t Q^{-1} r - r^t p - \frac{1}{2}[r^t Q^{-1} r - 2r^t Q^{-1} Q p + p^t Q p]\\ &= \frac{1}{2}(r^t Q^{-1} r - p^t Q p),\end{aligned} \tag{9.74}$$

hence

$$\begin{aligned}\frac{2}{\lambda}\phi = \beta V^2 \hat{1}^t w \left[\frac{1}{1+\gamma} w^{-1} + \frac{\gamma}{1+\gamma} S\right] w \hat{1}\\ - \beta p^t w[(1+\gamma)I - \gamma S w]p\end{aligned} \tag{9.75}$$

and

$$\begin{aligned}\frac{2}{\beta\lambda}\phi &= V^2\left[\frac{1}{1+\gamma} 1^t w \hat{1} + \frac{\gamma}{1+\gamma} 1^t w \hat{1} \hat{1}^t w \hat{1}\right] - p^t w[(1+\gamma)I - \gamma S w]p\\ &= V^2\left[\frac{1}{1+\gamma} + \frac{\gamma}{1+\gamma}\right] - (1+\gamma)\sum_{j=1}^{n} w_j p_j^2 + \gamma(p^t w \hat{1})(\hat{1}^t w p)\\ &= V^2 - (1+\gamma)\sum_{j=1}^{n} w_j p_j^2 + \gamma\left(\sum_{j=1}^{n} w_j p_j\right]^2\\ &= V^2 - (1+\gamma)(\sigma_p^2 - \bar{p}^2) + \gamma\bar{p}^2\\ &= V^2 - \bar{p}^2 - (1+\gamma)\sigma_p^2.\end{aligned}$$

Thus we may write the utility function in terms of price as

$$\boxed{\phi(p) = \frac{\lambda\beta}{2} V^2 - (1+\gamma)\sigma_p^2} \tag{9.76}$$

where σ_p^2 is the weighted variance, which equals $(\Sigma w_j p_j^2) - (\Sigma w_j p_j)^2$ and $\bar{p}$ is the weighted average, which equals $\Sigma w_j p_j$. Consumer surplus is given by

$$S = \phi - \lambda p^T q, \tag{9.77}$$

where $q = \beta w(V\hat{1} - [(1+\gamma)I - \gamma S w]p)$. Total surplus or gain to all parties is

$$\begin{aligned} T &= S + \Pi - S + (p - c)^T q = \phi - c^T q \\ &= \phi - c^T \beta w(V\hat{1} - [(1 + \gamma)I - \gamma S w]p) \\ &= \phi - \beta[Vc - (1 + \gamma)\Sigma w_i c_i p_i + \gamma \bar{c}\bar{p}] \\ &= \phi - \beta[V\bar{c} - (1 + \gamma)(\sigma_{cp}^2 + \bar{c}\bar{p}) + \gamma \bar{c}\bar{p}] \\ &= \phi - \beta[V\bar{c} - \bar{c}\bar{p} - (1 + \gamma)\sigma_{cp}^2] \\ &= \frac{\beta}{2}[V^2 - \bar{p}^2 - 2V\bar{c} + \bar{c}\bar{p} - (1 + \gamma)(\sigma_p^2 - 2\sigma_{cp}^2)], \end{aligned} \tag{9.78}$$

where σ_{cp} is the covariance of price and cost.

In particular for any set of prices the departure from efficiency is measured by

$$\begin{aligned} T(c) - T(p) &= \frac{\beta}{2}[\bar{p}^2 + \bar{c}^2 + 2\bar{c}\bar{p} + (1 + \gamma)(\sigma_c^2 + \sigma_p^2 - 2\sigma_{cp}^2)] \\ &= \frac{\beta}{2}[(\bar{p} - \bar{c})^2 + (1 + \gamma)(\sigma_p^2 - 2\sigma_{cp}^2 - \sigma_c^2)]. \end{aligned} \tag{9.79}$$

In a later section we apply this measure calculating both profits and consumer surplus for market examples for the different solutions suggested in this chapter. We also select parameters that might at least crudely reflect a three-firm oligopolistic automobile market.

Cooperation Parameter Solution

Although the cooperation parameter solution was mentioned in Chapters 5 and 6 and used in a calculation in Chapter 6, we did not give a method for obtaining the values of the cooperation parameters. Assume that each firm wishes to maximize

$$\prod_i = P_i - b_i \sum_{j=1}^{n} P_j, \tag{9.80}$$

where b_i represents Player i's attitude toward the average payoffs of all others. In Chapters 5 and 6 a slightly different formulation than that of Eq. (9.80) was used. We considered

$$\prod_i = P_i + \Theta_i \sum_{j \neq i} P_j. \tag{9.81}$$

Equations (9.80) and (9.81) are equivalent if we set $\Theta_i = b_i/(b_i - 1)$. For subsequent analysis, Eq. (9.80) is more convenient to work with than Eq. (9.81). In this new form, when $b_i = -\infty$, Eq. (9.80) is equivalent to joint maximization. When $b_i = 0$, the noncooperative solution is obtained and when $b_i = 1/n$, Eq. (9.80) gives the beat-the-average solution.

The relation of the cooperation parameter model to the maximization of market share comes about in the obtaining of an algorithm based on treating the expression

$$\frac{P_i}{\Sigma P_j} = b_i(p_1, p_2, \ldots, p_n)$$

as a constant, solving the resultant model and then iterating with new values for b_i. We must first demonstrate that we can solve the model given by Eq. (9.80).

Differentiating Eq. (9.80) with respect to p_i we obtain

$$\frac{\partial P_i}{\partial p_i} = b_i \Sigma \frac{\partial P_i}{\partial p_i}. \tag{9.82}$$

for $i = j$

$$\begin{aligned}\frac{\partial P_i}{\partial p_i} &= \beta w_i[V - p_i - \gamma(p_i - \bar{p})] - (p_i - c_i)[1 + \gamma(1 - w_i)] \\ &= \beta w_i(V + c_i[1 + \gamma(1 - w_i)] - [2(1 + \gamma) - \gamma w_i]p_i + \gamma\bar{p}).\end{aligned} \tag{9.83}$$

For $i \neq j$,

$$\frac{\partial P_j}{\partial p_i} = \beta w_j(p_j - c_j)\gamma w_i. \tag{9.84}$$

Summing, we obtain

$$\begin{aligned}\frac{\partial P_j}{\partial p_i} &= \gamma\beta w_i \sum_{j \neq i} w_j(p_j - c_j) \\ &\quad + \beta w_i[\gamma w_i(p_i - c_i) + V + (1 + \gamma)c_i - 2(1 + \gamma)p_i + \gamma\bar{p}] \\ &= \gamma\beta w_i \Sigma w_j(p_j - c_j) + \beta[V + (1 + \gamma)c_i - 2(1 + \gamma)p_i + \gamma\bar{p}] \\ &= \gamma\beta w_i(\bar{p} - \bar{c}) + \beta[V + (1 + \gamma)c_i - 2(1 + \gamma)p_i + \gamma\bar{p}].\end{aligned} \tag{9.85}$$

Substituting this equation in Eq. 9.82 and factoring out βw_i yields

$$\begin{aligned}V + [(1 + \gamma) - \gamma w_i]c_i &= [2(1 + \gamma) - \gamma w_i]\gamma_i + \gamma\bar{p} \\ &= b_i[V + (1 + \gamma)c_i - 2(1 + \gamma)p_i + 2\gamma\bar{p} - \bar{c}].\end{aligned} \tag{9.86}$$

Writing Eq. (9.86) more generally in matrix form, collecting terms, and solving for b we have

$$\begin{aligned}b &= [2(1 + \gamma)(I - R) - \gamma W - \gamma(I - 2R)SW]^{-1} \\ &\quad \times [V(I - R)\hat{1} + [(1 + \gamma)(I - R) - \gamma(I - RS)W]C].\end{aligned} \tag{9.87}$$

Call the first term on the right-hand side of Eq. (9.87) X^{-1}, hence

$$X = \gamma(I - 2R)\left[(I - 2R)^{-1}\left(\frac{2(1 + \gamma)}{\gamma}(I - R) - W\right)W^{-1} - S\right]W. \tag{9.88}$$

From the general relation we have

$$(Y^{-1} - S)^{-1} = \left[Y + \left(\frac{1}{1 - \Sigma y_i}\right)YSY\right]. \tag{9.89}$$

Define

$$Y^{-1} = (I - 2R)^{-1} \left[\frac{2(1+\gamma)}{\gamma}(I - R) - W\right] W^{-1}; \tag{9.90}$$

then

$$X^{-1} = \frac{1}{\gamma} W^{-1} \left[Y + \frac{1}{1 - \Sigma y_i} YSY\right] (I - 2R)^{-1}. \tag{9.91}$$

Call

$$Z = \frac{2(1+\gamma)}{\gamma}(I - R) - W; \tag{9.92}$$

then

$$\begin{aligned} X^{-1} &= \frac{1}{\gamma} W^{-1} \Big[WZ^{-1}(I - 2R) \\ &\quad + \frac{1}{1 - 2\gamma} WZ^{-1}(I - 2R)SWZ^{-1}(I - 2R)\Big] (I - 2R)^{-1} \\ &= \frac{1}{\gamma} Z^{-1} + \frac{1}{\gamma(1 - \Sigma y_i)} Z^{-1}(I - 2R)SWZ^{-1}, \end{aligned} \tag{9.93}$$

where

$$z_i^{-1} = \frac{\gamma}{2(1+\gamma)(1 - b_i) - \gamma w_i}$$

and

$$y_i = \frac{\gamma w_i(1 - 2b_i)}{2(1+\gamma)(1 - b_i) - \gamma w_i}.$$

The typical element of X^{-1} may be written as

$$\begin{aligned} x_{i,j}^{-1} &= \frac{\delta_{ij}}{2(1+\gamma)(1 - b_i) - \gamma w_i} \\ &\quad + \left(\frac{1}{1 - \sum_k y_k}\right)\left(\frac{\gamma(1 - 2b_i)w_j}{[2(1+\gamma)(1 - b_i) - \gamma w_i][2(1+\gamma)(1 - b_j) - \gamma w_j]}\right), \end{aligned} \tag{9.94}$$

where δ_{ij} is the Kronecker delta.

Substituting X^{-1} in Eq. (9.87) we obtain the general solution for b. This does not have a particularly convenient form, but does provide a closed-form expression for the evaluation of any particular case.

We consider a special case with symmetric $W = (1/n)I$ and $R = bI$. This means that the firms are symmetric in market share and in their views of the profit of the market as a whole. This gives

$$y_i = \frac{\gamma(1-2b)/n}{2(1+\gamma)(1-b)-\gamma/n} = \frac{\gamma(1-2b)}{2n(1+\gamma)(1-b)-\gamma} \tag{9.95}$$

and

$$\frac{1}{1-\Sigma y_i} = \frac{1}{1-\dfrac{n\gamma(1-2b)}{2n(1+\gamma)(1-b)-\gamma}} = \frac{2n(1+\gamma)(1-b)-\gamma}{2n(1-b)+\gamma(n-1)}. \tag{9.96}$$

Thus

$$X^{-1} = \left(\frac{n}{2n(1+\gamma)(1-b)-\gamma}\right) I + \left(\frac{1}{1-\Sigma y_k}\right)\left(\frac{n(1-2b)\gamma}{[2n(1+\gamma)(1-b)-\gamma]^2}\right) S \tag{9.97}$$

or

$$X^{-1} = \left(\frac{n}{2n(1+\gamma)(1-b)-\gamma}\right) I + \left(\frac{n(1-2b)\gamma}{[2n(1-b)+\gamma(n-1)][2n(1+\gamma)(1-b)-\gamma]}\right) S. \tag{9.98}$$

Returning to the second term of Eq. (9.87), it may be written as

$$V(1-b)\hat{1} + \left[(1+\gamma)(1-b)c - \frac{\gamma}{n}c + \gamma b\bar{c}\hat{1}\right] = V(1-b)\hat{1} + \frac{1}{n}([n(1+\gamma)(1-b)-\gamma]c + \gamma bn\bar{c}\hat{1}). \tag{9.99}$$

The value of b may now be calculated:

$$b_i = \left[\frac{n}{2n(1+\gamma)(1-b)-\gamma}\right]\left[V(1-b)\left(1+\frac{(1-2b)\gamma n}{2n(1-b)+\gamma(n-1)}\right) + \left(\frac{n(1+\gamma)(1-b)-\gamma}{n}\right)c_i + \left(\frac{[n(1+\gamma)(1-b)-\gamma](1-2b)}{2n(1-b)+\gamma(n-1)}\right)\bar{c} + \gamma b\bar{c}\left(1+\frac{(1-2b)\gamma n}{2n(1-b)+\gamma(n-1)}\right)\right], \tag{9.100}$$

or

$$p_i = \left[\frac{n(1-b)V}{2n(1-b)+(n-1)\gamma}\right] + \left[\frac{n(1+\gamma)(1-b)-\gamma}{2n(1+\gamma)(1-b)-\gamma}\right]c_i + \left[\frac{n\gamma\bar{c}}{2n(1-b)+(n-1)\gamma}\right]\left[\frac{n+(n-1)\gamma}{2n(1+\gamma)(1-b)-\gamma}\right]. \tag{9.101}$$

Equation (9.101) can be rewritten as

$$p_i = \frac{nV}{2n + \frac{n-1}{1-b}\gamma} + \frac{1}{2n(1+\gamma) - \frac{\gamma}{1-b}} \times \left[\left(n(1+\gamma) - \frac{\gamma}{1-b}\right) c_i + \left(\frac{n\gamma(n + (n-1)\gamma)\bar{c}}{2n(1-b) + (n-1)\gamma}\right)\right]. \quad (9.102)$$

We may verify the special cases

$$b = \begin{cases} -\infty & \text{or joint maximum} \\ 0 & \text{or noncooperative equilibrium} \\ 1/n & \text{or beat-the-average.} \end{cases}$$

As $b \to -\infty$, we have

$$p_i \to \frac{nV}{2n} + \frac{1}{2n(1+\gamma)}[n(1+\gamma)c_i + 0] = \frac{1}{2}(V + c_i). \quad (9.103)$$

For $b = 0$, we have

$$\begin{aligned} p_i &= \frac{nV}{2n + (n-1)\gamma} \\ &\quad + \frac{1}{2n(1+\gamma) - \gamma}\left([n(1+\gamma) - \gamma]c_i + \frac{n\gamma[n + (n-1)\gamma]\bar{c}}{2n + (n-1)\gamma}\right) \\ &= \frac{V}{2 + \frac{n-1}{n}\gamma} + \left(\frac{1 + \frac{n-1}{n}\gamma}{2(1+\gamma) - \frac{\gamma}{n}}\right)\left(c_i + \frac{\gamma\bar{c}}{2 + \frac{n-1}{n}\gamma}\right) \end{aligned} \quad (9.104)$$

which is the noncooperative equilibrium for nonsymmetric costs. Finally, setting $b = 1/n$, we obtain

$$p_i = \frac{nV}{2n + \frac{n-1}{1 - \frac{1}{n}}\gamma} + \frac{1}{2n(1+\gamma) - \frac{n\gamma}{n-1}}(K) \quad \text{where}$$

$$\begin{aligned} K &= \left(\left[n(1+\gamma) - \frac{n}{n-1}\gamma\right] c_i + \frac{n\gamma[n + (n-1)]\gamma\bar{c}}{2(n-1) + (n-1)}\right) \\ &= \frac{V}{2+\gamma} \\ &\quad + \frac{1}{2(n-1)(1+\gamma) - \gamma}\left([(n-1)(1+\gamma) - \gamma]c_i + \frac{[n + (n-1)\gamma]\bar{c}}{2+\gamma}\right). \end{aligned}$$

Setting $\bar{c} = c_i = c$, we obtain

$$p_i = \frac{1}{2+\gamma}[V + (1+\gamma)\bar{c}].$$

The solution of the cooperation parameter model gives the necessary computation method for the iteration procedure to calculate the maximize-profit-share solution and also the maximize-market-share solution.

Miscalculation and Inefficiency

In this chapter we have investigated a battery of solutions and are able to calculate the prices, outputs, and profits associated with them. If we wish to examine data arising from the playing of an oligopoly game or from an actual market, we must take into account several added features of both structure and behavior before we can interpret the outcomes in terms of our theorizing.

If we are using empirically estimated parameters, the comparative analysis of solutions and consumer surplus may easily be suspect since our model of market structure uses a linear approximation of industry demand. This may be reasonable in the zone for which we have information but may not hold all the way between the monopoly price and competitive equilibrium.

In some instances profits may be high because a series of blunders cancel themselves out. The market forecast may be wrong on the high side, but fortunately a production blunder is committed which results in a failure to produce for the projected demand but adequate production for the actual demand.

To evaluate adequately the performance of a firm, individual measures are needed to estimate market structure; demand (this includes the game theoretic problem of taking into account competitors' behavior), and the measurement of the success of production and inventory scheduling, given the market forecasts. Measures are also needed for success in financial and capacity investment policies. Finally a measure is needed for success in coordinating these separate activities. A favorable balance sheet is not an adequate measure of success. As in billiards, so in the market place, it is one thing to call your shots before the event and another to spend time rationalizing a chance success.

Nonsymmetric Oligopoly Game and Market Models

Parameters and Solutions: Two- and Three-Firm Examples

Four specific examples of markets are presented in this section. Different solutions for each are calculated and displayed in the text. The first is based on crude figures reflecting highly aggregated information on the automobile market. The other three are duopoly models that should be compared with those found in Chapters 5 and 6.

Before presenting the calculations, we wish to summarize our approach and purpose. We believe that a nonsymmetric oligopoly model based on demand derived from an aggregate quadratic utility function provides us with the possibility of obtaining explicit solutions, estimating parameters, and constructing a manageable simulation of the market that can be used for an oligopoly game.

We believe that the benchmark of the price noncooperative equilibrium is as natural to the examination of oligopolistic structures as is the competitive equilibrium for competitive markets. In fact we believe that in general the noncooperative equilibrium is a far more basic and general concept than the competitive equilibrium. The latter is merely a highly special case of the former.

No pretense is made in this section to offer new or original econometrics. We stress the possibility of constructing enough structure to enable use of a battery of static oligopoly results in conjunction with econometric estimation.

A Simulation of an Automobile Market If the automobile industry were in noncooperative equilibrium we could use the noncooperative solution together with industry information to calculate parameter estimates for the automobile industry. A very crude set of parameters estimates were obtained from the 1965 figures for the three major automobile companies. The calculations given are merely meant to be suggestive of an approach and not a careful econometric estimate of the automobile industry. The game constructed will be somewhat like the industry.

Owing to the lack of unconsolidated figures, several approximations and simplifications were made. In particular we consider only those firms and their worldwide competition. We speak generically of all automotive units, such as autos, trucks, and tractors. We know that civilian nonauto products and defense products accounted for $1.9 billion for General Motors or approximately 10% of sales. Rather than break out the multiproduct features explicitly we implicitly include them by inflating the price of an automotive unit so that we make the crude approximation that there is a constant ratio in multiproduct sales. Furthermore the distribution system is not accounted for explicitly. The firms obtain wholesale prices but the cars are sold at retail.

The following tabulation gives sales (not corrected for total income which is slightly different), total assets, and before tax profits:

AUTO INDUSTRY	SALES ($\times 10^6$)	ASSETS ($\times 10^6$)	PROFITS ($\times 10^6$)
General Motors	20,734	11,479	2,126
Ford	11,537	7,596	710
Chrysler	5,300	2,934	233

Variable costs are assumed to correspond to costs of products sold on the earnings statements in the annual reports. Depreciation, amortization, administrative expenses, debt servicing, and various pension and retirement payments are assumed to define fixed costs. Taxes are reported separately, they include foreign taxes.

Auto industry	Variable costs ($\times 10^6$)	Fixed costs ($\times 10^6$)	Taxes ($\times 10^6$)
General Motors	15,250	1,559	1,966
Ford	8,853	1,401	596
Chrysler	4,121	746	213

A crude indication of the physical sizes of the corporations is given by the value placed on plant, equipment, property, and special tools. These figures, of course, are highly influenced by accounting practices and may grossly underestimate the worth of capital investment especially when land values have increased.

Auto industry	Property and plant ($\times 10^6$)	Special machinery ($\times 10^6$)
General Motors	4,161	455
Ford	2,574	446
Chrysler	887	180

World sales of automotive equipment including trucks and tractors is given in the following table:

Auto industry	Sales ($\times 10^3$)	%
General Motors	7,278	52.2
Ford	4,595	32.9
Chrysler	2,077	14.9
Total	13,950	100.0

On the basis of the assumption of linear costs from the above information we may write

$$p_1 = 20{,}734/7.278 = 2{,}849 \qquad c_1 = 15{,}250/7.278 = 2{,}095$$
$$p_2 = 11{,}537/4.595 = 2{,}511 \qquad c_2 = 8{,}853/4.595 = 1{,}927$$
$$p_3 = 5{,}300/2.077 = 2{,}552 \qquad c_3 = 4{,}121/2.077 = 1{,}984.$$

We assume that the demand for the automotive products of any firm i is given by

$$q_i = \beta w_i[V - p_i - \gamma(p_i - \Sigma w_j p_j)], \quad \text{for} \quad i = 1, 2, 3. \tag{9.105}$$

From Eq. (9.26) we have

$$p_i = \frac{1}{\Delta - w_i}\left(\frac{V}{\gamma}Q + \frac{1}{2}[(\Delta - 2w_i)c_i + Q\Sigma z_j(\Delta - 2w_j)q]\right). \tag{9.106}$$

We know that $w_1 + w_2 + w_3 = 1$ or $w_3 = 1 - w_1 - w_2$; thus we have five undetermined parameters β, α ($V = \alpha/\beta$ hence we determine α rather than V), γ, w_1, and w_2.

In Eq. (9.106), $V = \alpha/\beta$, $\Delta = 2(1 + \gamma)/\gamma$, $z_i = 1/x_i$, $x_i = \Delta(1/w_i) - 1$, and $Q = 1/[1 - \Sigma_j z_j]$. From the three equations of the form

$$q_i = \beta w_i[V - (1 + \gamma)p_i + \gamma \Sigma p_j w_j], \qquad (9.107)$$

we obtain by subtraction

$$\frac{q_1/w_1 + q_2/w_2}{p_2 - p_1} = \frac{q_1/w_1 - q_2/w_3}{p_3 - p_1} = (1 + \gamma)\beta, \qquad (9.108)$$

from which we derive

$$w_2 w_3 q_1(p_3 - p_2) + w_1 w_3 q_2(p_1 - p_3) + w_1 w_2 q_3(p_2 - p_1) = 0. \qquad (9.109)$$

Let $q_i(p_j - p_k) = z_i$; we may rewrite Eq. (9.109) as

$$z_1 w_2 w_3 + z_2 w_1 w_3 + z_3 w_1 w_2 = 0. \qquad (9.110)$$

We know that $w_1 + w_2 + w_3 = 1$; hence

$$w_3 = 1 - w_1 - w_2. \qquad (9.111)$$

Substituting in Eq. (9.110), we obtain

$$z_1 w_2(1 - w_1 - w_2) + z_2 w_1(1 - w_1 - w_2) + z_3 w_1 w_2 = 0, \qquad (9.112)$$

which yields

$$-z_2 w_1^2 - z_1 w_2^2 + [z_1(1 - w_1) - z_2 w_1 + z_3 w_1] w_2 + z_2 w_1 = 0. \qquad (9.113)$$

Dividing Eq. (9.113) by $-z_1$, we obtain

$$w_2^2 - \left[1 + \frac{(z_3 - z_1 - z_2)}{z_1} w_1\right] w_2 - \frac{z_2}{z_1} w_1(1 - w_1) = 0. \qquad (9.114)$$

Setting

$$\frac{z_3 - z_1 - z_2}{z_1} = r_1, \quad \frac{z_2}{z_1} = r_2,$$

we may rewrite Eq. (9.114) as

$$w_2^2 - (1 + r_1 w_1) w_2 - r_2 w_1(1 - w_1) = 0. \qquad (9.115)$$

We solve this equation to obtain w_2 as a function of w_1. We then search through successive values of w_1 until we obtain a positive root. Returning to Eq. (9.109) we may express β as a function of w_1 and γ,

$$\beta(w_1, \gamma) = \frac{q_1/w_1 - q_2/w_2}{(p_2 - p_1)(1 + \gamma)}. \qquad (9.116)$$

Using this equation in Eq. (9.109) to eliminate β, we solve for V:

$$V(w_1, \gamma) = \frac{q_1}{\beta w_1} + (1 + \gamma)p_1 - \gamma \Sigma p_j w_j. \tag{9.117}$$

This now leaves us the problem of estimating w_1 and γ. We do this using the Chebychef Criterion, which involves minimizing the maximum of the absolute value of the ratio of the deviation of predicted from observed prices.

Using the crude aggregated data for p_i and c_i obtained from the yearly reports, we have, for $p_1 = 2{,}849$,

$$V = 533{,}678 \quad \beta = 26.3 \quad \gamma = 1.988$$
$$w_1 = .75 \quad w_2 = .17 \quad w_3 = .08$$
$$\text{maximum deviation} = 0.0080$$

These estimates appear to be somewhat startling as can be seen from β, which implies that a \$100 cut from all firms would result in the sale of 2,600 more automobiles! We noted previously, however, that the aggregation used to obtain the observed average prices does not appear to be reasonable owing to the multiproduct nature of the firms. General Motors especially has an important part of its business (and hence costs and sales) in markets other than vehicles. We suspect that a more detailed gathering of statistics would somewhat lessen the differences in observed aggregate average prices which apparently have General Motors prices (and costs) considerably above the others. We reduce this difference (of approximately \$325) by \$50 and\$100 to view the effect on our parameter estimation. Setting $p_1 = \$2{,}799$ and then $p_1 = \$2{,}749$, we obtain, for $p_1 = 2{,}799$,

$$V = 6{,}039.3 \quad \beta = 4{,}198.5 \quad \gamma = 9.28$$
$$w_1 = .70 \quad w_2 = .20 \quad w_3 = .10$$
$$\text{maximum deviation} = .0085$$

for $p_1 = 2{,}749$,

$$V = 3{,}951 \quad \beta = 10{,}889 \quad \gamma = 2.2$$
$$w_1 = .65 \quad w_2 = .24 \quad w_3 = .11$$
$$\text{maximum deviation} = .0124$$

A quick crude check of the above values shows that when $p_1 = 2{,}799$,

$$e = \frac{p\Delta q}{q\Delta p} \simeq .84,$$

which seems to be somewhat low. When $p_1 = 2{,}749$,

$$e \simeq 2.1;$$

Table 9.1 Parameters for the automobile industry.

AUTO INDUSTRY	WEIGHTS (w_i)	AVERAGE COSTS
General Motors	.65	1870[a]
Ford	.236	1927
Chrysler	.114	1984

[a] Our crude calculations give the average costs of General Motors higher than the others. We believe this to be due to the crudeness of our aggregation (2,095). For this calculation we reduce General Motors' costs.

this appears to be somewhat high. It is evident that for specific econometric use of our method we need better statistics on prices and costs than the ones we have used. However by means of a computer program, which is not given here, the joint maximum, price noncooperative equilibrium, efficient prices, and consumers' surplus are calculated using slightly different estimates as shown in Table 9.1. The results are displayed in Table 9.2. The p_i are prices, P_i are net variable revenues (no overheads are subtracted), and Φ is the consumer utility.

We make no pretense at accuracy; calculations are offered only to suggest the relative sizes of the different solutions. For example it appears that the actual market is very close to and slightly more inefficient than the price noncooperative equilibrium.

The relative efficiencies of the solutions are approximately in the ratio of 100:94:94.5:75 ranging from the efficient solution to total cooperation by the firms. The loss to the public is, of course, greatest when the firms collude.

It is of interest to note that for n firms we would have V, β, γ, and $w_1, w_2, \ldots, w_{n-1}$ to estimate or $n + 2$ parameters in toto. We have $2n$ equations that must be satisfied. In the case of the automobile market examined here we have five parameters and six equations. Our computational methods can be extended for more than three firms at the cost of some complication in the ease of manipulation.

Table 9.2 Solutions for the automobile industry as a triopoly.

SOLUTION	p_1	p_2	p_3	P_1	P_2	P_3	Φ	TOTAL SURPLUS
Joint maximum	3486	3515	3543	10156	3272	1394	39730	22233
Actual	2767	2511	2552	6528	2683	1180	54077	27492
Noncooperative equilibrium	2703	2572	2565	6813	2560	1133	54525	27813
Efficient point	1870	1927	1984	0	0	0	64637	29644

SOME DUOPOLY EXAMPLES

Three examples of nonsymmetric duopolies are given:

All cases: $\alpha = 2 \times 10^6$, $\beta = 2{,}500$, $n = 2$, $\gamma = 3$.
Case 1: $w_1 = w_2 = 0.5$, $c_1 = 250$, $c_2 = 200$.
Case 2: $w_1 = 0.75$, $w_2 = 0.25$, $c_1 = c_2 = 200$.
Case 3: $w_1 = 0.75$, $w_2 = 0.25$, $c_1 = 250$, $c_2 = 200$.

The profit diagrams in Figure 9.2 show the solutions given in Table 9.3. The nonsymmetry of the three games is indicated by the position of the various solution points. These should be compared with Figure 6.5. The Pareto optimal surface will differ in each case.

On Static Oligopoly Theory

This volume has been devoted primarily to the detailed exploration of a class of simple oligopoly models stressing the examination of structure. Static or equilibrium solutions have been considered even though it is evident that many of the key problems in oligopoly theory entail a study of process.

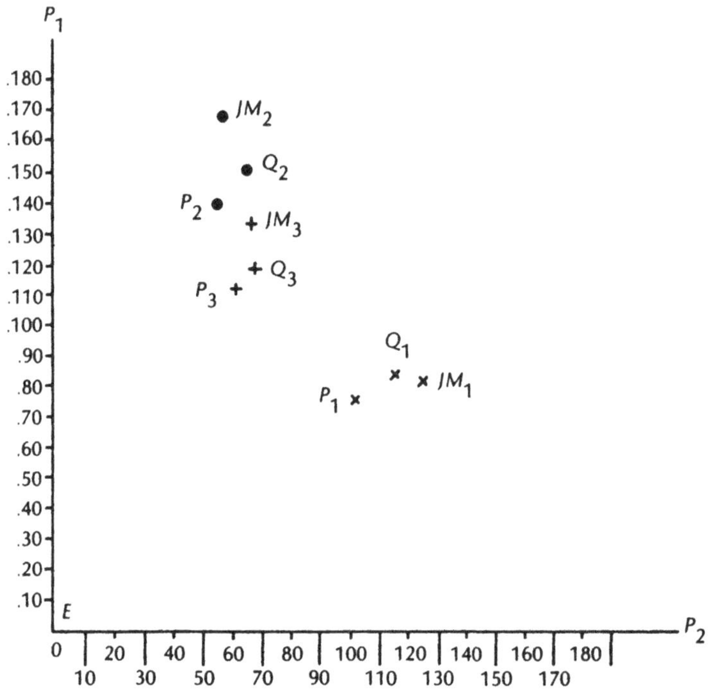

Figure 9.2. Profits of nonsymmetric duopolists.

Table 9.3 Solutions for nonsymmetric duopoly games.

SOLUTION	p_1	p_2	q_1 ×1,000	q_2 ×1,000	P_1 ×10[6]	P_2 ×10[6]
CASE 1						
Joint maximum	525.00	500.00	296.88	421.88	81.64	126.56
Noncooperative equilibrium						
Quantity	453.30	439.01	406.59	478.02	82.66	114.25
Price	398.90	379.67	465.31	561.47	69.29	100.88
Efficient point	250.00	200.00	593.75	843.75	0	0
CASE 2						
Joint maximum	500.00	500.00	562.50	187.50	168.75	56.25
Noncooperative equilibrium						
Quantity	454.60	411.87	587.54	302.67	149.59	64.13
Price	406.53	363.80	677.67	332.72	139.96	54.50
Efficient point	200.00	200.00	1,125.00	375.00	0	0
CASE 3						
Joint maximum	525.00	500.00	480.47	222.66	132.13	66.80
Noncooperative equilibrium						
Quantity	447.60	421.22	525.22	316.02	119.54	69.91
Price	433.53	373.15	602.21	351.70	110.52	60.90
Efficient point	250.00	200.00	960.94	445.31	0	0

This approach is defended with the observation that before one can study process in any depth it is desirable to understand the underlying mechanisms that carry the process. These are the markets and firms and other commercial institutions of society. At the highest level of abstraction the actors and institutions are defined by the *rules of the game*. Thus the Cournot model of duopoly, for example, is a full description of the firms and the type of markets in which they exist. The model itself states nothing about equilibrium or disequilibrium. The solution concept of noncooperative equilibrium which is usually applied to the Cournot model enables the economist to select an outcome consistent with the intents of the firms. There is no dynamic or behavioral description supplied.

With the exception of a few highly simplified dynamic mathematical models of oligopoly, the tradition in dealing with the problems of competition in a dynamic context has called for a mixture of verbal and diagrammatic methods. These models are open, dealing with partial equilibrium or disequilibrium. They are implicitly or explicitly institutional and they describe trade and production utilizing money. Much of this work has been covered in Chapter 3.

In contrast with the verbal approach to dynamics, there is work on mathematical models of partial equilibrium in oligopolistic markets, such as developed here, and mathematical models of general equilibrium. It is natural to ask whether these different approaches can be reconciled. For what purposes should these approaches be reconciled and what can we expect to learn in doing so?

The work of Cournot, the mathematical models of oligopolistic markets presented here, and the many others referred to are generally full descriptions of firms and markets in which specific institutions (firms) sell goods and services to an outside world (represented by given demand conditions) and obtain money. The institutional aspects even of the mathematical models come in features such as U-shaped average-cost curves. Implicitly packed into the story of decreasing, followed by flat, followed by rising average costs are considerations of organization, overheads, and capacity.

Even with simple open models the mathematically general proof of the existence of a noncooperative equilibrium calls for conditions on costs. But certainly as long as the production possibility sets are convex (increasing returns to scale ruled out) and enough conditions are placed on the structure of demand, noncooperative equilibria can be shown to exist. (We return to this point below.)

The general equilibrium theory of Walras (1954) was formally mathematized in a totally noninstitutional manner by Arrow and Debreu (1954) and Debreu (1959). In these mathematical models there is really only one economic actor, the consumer resource-holder trader and a set of firms run by shadows or automata in the form of profit maximizing managers whose only goal is to flow through 100% of their profits to the stockholders.

The existence of a price system (not necessarily unique) that brings about Pareto optimal trade is established and it is shown that all books balance at equilibrium. There is no need to postulate the existence of money in this model. If we wish, some item may have its price set at 1 and hence serve as a numeraire.

The elegance and generality of the Arrow-Debreu model of general equilibrium makes one raise the question about reconciling the mathematical models of oligopoly theory with it for several reasons. If accomplished, all feedbacks, actions, and reactions excluded from the partial equilibrium models would be accounted for. Furthermore, a general existence proof for a noncooperative equilibrium would represent a considerable generalization of the scope of most of the current models. If the equilibrium were not at a Pareto optimal point this would establish with full rigor the failure to achieve efficiency under oligopolistic competition. If no noncooperative equilibrium exists (as in the Edgeworth example, or as we have maintained here in many instances in the Chamberlin model—see Chapter 8), we must question the models, the solution concepts, and the empirical evidence.

If we could embed the open oligopoly models that are frequently solved for noncooperative equilibria into a general equilibrium framework it would be possible to use the techniques of replication (or alternatively of "chopping up" decision units) to see if oligopolistic competition approaches competitive behavior. The replication used in Chapter 6 lends credence to these conclusions.

Monopolistic Competition in a Closed Economic System

In the course of the last forty years much terminological heat and little light has been produced in distinguishing among imperfect, monopolistic, and oligopolistic competition. In this work we note that the two items of concern are the closeness of strategic interlinkage between the products of any two firms and the number of firms. On this basis competition among few firms selling an identical product is oligopolistic competition. But we can regard this as a limiting case of monopolistic competition. Thus the phrases monopolistic or oligopolistic competition are used interchangeably.

Another source of confusion in terminology is the use of the phrase general equilibrium model rather than closed economic model. The general equilibrium analysis of Arrow and Debreu is applied to their closed economic model of trade and production. The phrase general equilibrium model leads to confusion between the economic model and the solution concept and analysis applied to the model. Here the phrase closed economic model is used to mean an economic model in which all feedbacks are accounted for.

In the past few years, various individuals (Negishi, 1961; Arrow and Hahn, 1971; Gabszewicz and Vial, 1972; Fitzroy, 1974; Marshak and Selten, 1974; Nikaido, 1975; Laffont and Laroque, 1976; Roberts and Sonnenschein, 1977) have attempted to embed monopolistic competition into a closed economic model and to relate the resulting analysis to the general equilibrium analysis. In these citations, with the exception of the last, the authors are able to establish the existence of a noncooperative equilibrium for the specific model being studied by means of special assumptions, such as assuming that consumers are pricetakers and that the payoffs to the firms have certain regular properties. Roberts and Sonnenschein (1977) criticize these authors for ad hoc assumptions arguing that if one makes use of the fundamental data of preferences, endowments, and technology as is done in general equilibrium theory it is possible to show that the noncooperative equilibrium (in pure strategies) may not exist. They construct a model of oligopolistic competition in a closed economy with Cournot or quantity setting firms demonstrating that their model has no perfect noncooperative equilibrium.

Shubik (1973a), Shapley and Shubik (1977), Shapley (1976), and Dubey and Shubik (1977a; 1978a) have also considered the problem of embedding monopolistic competition into a closed economic model. They have argued that the problem of building an adequate model of oligopolistic competition in a closed economy depends delicately on careful specification of information conditions and the strategies available to firms and customers.

Dubey and Shubik (1978b) have shown that the criticism of Roberts and Sonnenschein of attempts to embed models of oligopolistic competition within a general equilibrium context is misplaced. The criticism does not depend upon open or closed models, that is, has nothing to do with general equilibrium. The examples studied in Chapters 5 and 8 in this book based upon the work of Edgeworth are sufficient to show that the type of equilibrium sought by Roberts and Sonnenschein may not exist even in the partial equilibrium context.[1] However it is always possible to construct models of oligopolistic competition in a closed economy with pure strategy noncooperative equilibria using the same type of basic economic assumptions as are made in the study of general equilibrium although perfect equilibria do not exist.

The results they obtain are not sensitive to information conditions in one sense, yet are considerably sensitive in another sense. In particular they are not sensitive as a basic theorem (Dubey and Shubik, 1979) linking information conditions and equilibrium points shows that if two games (not involving exogenous uncertainty) differ only in information conditions such that one game has more information than the other then the noncooperative equilibrium points of the game with less information will always also be equilibrium points of the game with more information. The results are sensitive in that the introduction of more information into a game may add many new equilibrium points (frequently enforced by threat strategies which now become possible) or may add no equilibrium points.

We do not develop the technical apparatus needed to prove these results here. The interested reader may see them in the papers noted. However, we agree with the first observations of Roberts and Sonnenschein that one does not wish to make ad hoc restrictive assumptions on the models of oligopolistic competition. Without making any restrictive assumptions of the variety noted we are able to show that a noncooperative

1. The key distinction is the model of the consumer's set of strategies. Does he name a function as his strategy? That is, a complete excess demand schedule or is he constrained to a far simpler strategy?

equilibrium point always exists for oligopolistic markets (Dubey and Shubik, 1978b) with sufficiently little information and simple strategies. This proves the existence in a general way for the closed model, the open model, and all models differing only in the addition of information.

It is also shown that when the number of competitors is large (a nonatomic game is considered), the noncooperative equilibrium points coincide in prices and distribution of goods with the competitive equilibria of the closed economic model. Thus we do have consistency between the monopolistic competition analysis and the general equilibrium analysis.

Although it may be comforting to some to know that two highly, differently motivated sets of models can be made consistent and that results concerning noncooperative equilibria are validated, the thrust of the criticism of Shubik (1959b), Selten (1973), Roberts and Sonnenschein (1977), and earlier writers (Fellner, 1949; Stackelberg, 1952) still holds. With only a few competitors in a market, the strategies characterizing the Cournot, Edgeworth, general equilibrium models are too simple. When the competitors use more complicated strategies based on detailed use of information, no equilibria or many different equilibria may exist that are based on the more complex strategies.

The modeling needed to clarify conditions of strategic economic behavior, especially where information conditions and the sequencing of actions must be noted with care, calls for the explicit specification of the rules of money and other financial instruments. The reason is that money and financial instruments serve as important *strategic decoupling devices*. In equilibrium the books may balance, but at the decision point when individuals are buying and selling, paying out or receiving money, and shipping or receiving goods, there is no guarantee that the books are going to balance.

In disequilibrium imperfect balance is swept into the next period with the aid of financial instruments. In some circumstances in which credit outstanding is due but cannot be repaid, liquidation, reorganization, or bankruptcy are used to balance the books.

The strategic decisions of buying and selling are to a great extent financial decisions made in an ongoing process with no guarantee of perfectly balanced books. In most models the importance or need for a monetary and credit mechanism is difficult to see at an equilibrium point. The financial mechanism primarily facilitates decoupled decisionmaking away from equilibrium. But it is at the point of equilibrium that the books balance and the need for the financial mechanism apparently disappears.

It is suggested here that the eventual development of a satisfactory theory of oligopoly is intimately related to the development of a theory of money, financial instruments, and institutions. Much of the weaponry in economic competition is financial. In partial equilibrium studies this is implicitly, if not explicitly, acknowledged inasmuch as money plays a natural role.

Concluding Remarks

We have not dealt adequately with entry and exit particularly in a closed economic model or with the models of the firm and its management. One of these topics is dealt with in a separate paper (Dubey and Shubik, 1978c).

Entry and exit require a more fully dynamic context than our analysis has provided here. To best portray these conditions in a static closed economic model is to imagine an economy with individually owned technologies in which there are many alternative ways of manufacturing similar items that will be more or less efficient under various price structures. All individuals or firms are potential entrants in any industry for which they own a technology. The impact of entry can be studied via comparative statics. We may change some aspect of tastes or technology and compare the new with the previous equilibrium points.

We may consider the manager of an individual firm not as an automatic one-hundred percent fiduciary but as a goal-oriented or utility-maximizing individual who may or may not own shares in the firm he manages. It can then be shown that extra rules must be introduced to prevent blatant self-serving such as selling below market to other firms in which the manager may hold a major interest. Even with extra rules the manager will not necessarily maximize the profits of the firm unless it is relatively small with respect to all markets.

10 | Marketing, Advertising, and Product Differentiation

MARKETING, DISTRIBUTION, AND THEIR RELATED ACTIVITIES have been the stepchildren of economic theory. Despite the observations of Chamberlin, there has been little effort to integrate microeconomic theorizing and price theory with marketing studies until very recently. One can read a textbook in microeconomics and emerge with the notion that the economy is comprised of a group of manufacturing firms that offer their product directly to the final consumer. This assumption could not be further from the facts. Figure 10.1 shows the numbers of manufacturers, wholesalers, and retailers in the United States in 1974. We see from this figure that wholesalers and retailers far outnumber manufacturers. Not represented are the hosts of professionals such as doctors, dentists, or consulting engineers, who while not directly involved in the flow of goods guide the direction of the flow from manufacturers, wholesale or retail outlets to the final customer.

A product may be guided to the final customer from the manufacturer through anywhere from zero to three intermediaries. Very often the pattern of flow of goods is from a manufacturer to a distributor to a retailer to the public. Even services do not necessarily reach the ultimate consumer directly from the producer. Actors have agents; plays are staged in theaters whose management sells tickets; surgeons and other specialists are often reached through general practitioners.

Figure 10.2 illustrates the flow of two typical items to the consumer. Different firms in the same industry or, more precisely, selling closely re-

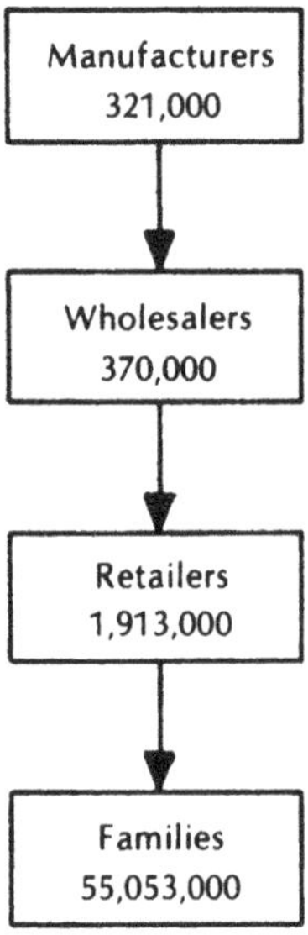

Figure 10.1. Comparative number of firms in the distribution chain (from *Statistical Abstract of the United States* (1976), 40, estimate for 1972).

lated products may easily have considerably different distribution and retailing systems. For example, one firm may be a pure manufacturer selling to independent wholesalers who are in turn selling to independent retailers who sell to the final customer. In some instances, firms own their own distributors. Other firms own both distributors and retailers or only have one distribution retailing organization. These varieties make for different levels of flexibility and give rise to different maneuverings in the market. For example, the main competition may not be among manufacturers fighting for the ultimate customer but may occur at an earlier stage during which the manufacturers fight for retail outlets.

It is evident that the nature of the strategic freedom of a manufacturing firm will depend on a host of specific variables such as the size of the product, the grading of the product, the frequency with which it is bought, the transportation difficulties involved in getting the product to the market, the consumer knowledge of the technical properties of the product, and many other features. Without ad hoc empirical investigations of relevant variables in the different markets, it would be presumptuous and misleading to suggest it possible to present a general theory that has more than a tenuous connection with the nature of competition in specific markets.

Despite Chamberlin's sensitive perceptions of the importance of marketing and product variation, he provided little more toward integrating marketing, distribution, and advertising with the study of monopolistic or oligopolistic competition beyond a few cogent observations. The obser-

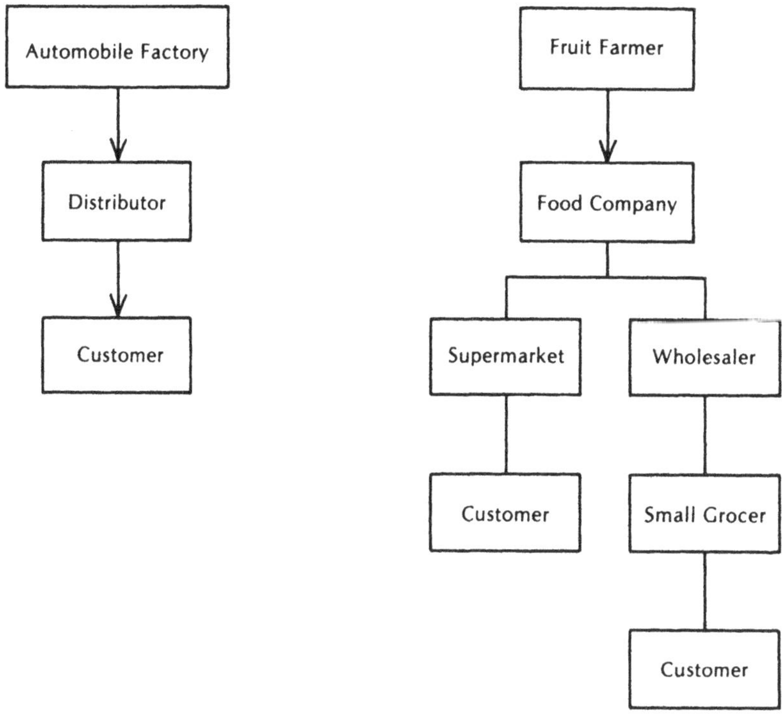

Figure 10.2. Distribution chains.

vation that a demand curve will shift with advertising expenditures is not particularly astute in and of itself. The more fundamental question is how will the curve shift, what proof exists that it will shift in such a manner, and for what class of items will such a shift occur. Analytical marketing literature attempting to answer some of these questions is not in great abundance. However, there are indications of change. The work of Bass and colleagues (1961) and Montgomery and Urban (1969) are harbingers of the concern for explicit models and the construction of measures in marketing.

How are we to describe the nature of competition in a world of oligopolies and product variation? How can we construct a relevant, logically consistent, and testable theory? When is a so-called theoretical result worth obtaining? The result that advertising shifts the demand curve tells us virtually nothing. If it asserts that it shifts the demand curve to the right, then the assertion could conceivably be proved or disproved and does narrow the field. However, there is little substantive or theoretical value to the statement that advertising shifts demand unless some indication is given as to how demand is shifted and what are the conditions of the shift.

In this and the next chapter we devote our time to sketching out the economic problems posed by distribution, retailing, advertising, and related topics. Our description and discussion must be interpreted as little more than a program and sketch for work that must be done if we are to presume that the tools of economic analysis can be brought to bear on specific problems in competition. In Chapter 11 we narrow our scope and look at several possibilities in the nature of shifts in demand and carry through an analysis of them.

Before we turn to some of the details of distribution and related topics, a few figures may help us appreciate the role of marketing, advertising, and other product variation activities in the United States. In 1972 the estimate of the amount spent on advertising was $23.3 billion. Advertising agencies numbered 28,400 with total receipts of $10.6 billion. These figures are crude in the extreme. Undoubtedly one category shades into the other in a manner that makes it very difficult to draw fine distinctions. Many agencies not only help with advertising but handle public relations and in some cases packaging as well.

In 1972 the total amount spent on advertising in the United States was between 100 and 110 dollars per capita. The disposable consumer income during that year was $3,837. Expenditures were $3,510. According to Firestone (1967) Canadian figures were approximately one-half that of the United States.

It is difficult to characterize the intensity of the change in markets and products in a dynamic society. It is frequently hard to define when a product is truly new. A crude estimate of innovative activity is at least supplied by expenditures on research and development. In 1969 total R&D was reported at 26 billion dollars (*Statistical Abstracts of the United States*, 1971, p. 509) of which 45% was defense-space oriented and less than 10 billion dollars was industry funded. Gross National Product was $930 billion (S.A.U.S., 1971, p. 305), hence R&D was approximately 2.8% of GNP.

Yet one might argue from another point of view that much of the expenditures on education, institutional organization, and advertising are part of new product expenditures. For it is these expenditures that characterize the differentiation of a dynamic evolving economy from a timeless market with static immortal traders and firms.

The consumer is not always completely passive in the role as an information gatherer. However, his expenditures are minute in comparison with those of industry. One index is the level of sales of *Consumers' Report:* in 1977 its circulation was 2,150,000.

The Weapons in the Arsenal

While the observation that there are far more variables than price and the decision to produce a certain quantity of a given standardized good, it only gets us so far in our understanding of competition. The observation comes close to being vacuous to add that some of these extra variables move the cost curve and others move the demand curve; hence a good economist has only to set *marginal everything* equal to *marginal everything else,* and all will be well in the world.

Unfortunately we are confronted with at least three tasks, none of which has been completed to date. A reasonable and relevant taxonomy is needed to describe the weapons in the competitor's arsenal in a manner that conforms both with common and uncommon sense. We need to construct measures for the intensity of these variables. In some cases these measures will by necessity be highly complicated. At least a dimensional analysis must be performed; that is, by what units do we consider the phenomena characterized? The problem of measurement consists of far more than deciding whether a cardinal or an ordinal scale exists. While we cannot pursue this topic further in this volume, both elementary and advanced treatments exist elsewhere (Fishburn, 1974).

Given that we have reasonable categories and some form to measure the variables, the next task is to construct and validate behavioral theories of this much enlarged system of competition. Precisely at this point the marriage of advanced methods in economics, the behavioral sciences, and marketing must come together.

How are demand and cost curves moved, by which variables, and how long does it take to move these curves? What are the implications of the movements? Chamberlin (1962) suggested in his writing that most of the problems in this area were qualitative. A statement such as this is merely a way of stating that this problem is not going to be analyzed in depth at this time. Chamberlin's contribution was large and undoubtedly he had neither the time nor the information to do very much more than he did concerning these particular problems. Nevertheless, depth of analysis is called for if microeconomic analysis is worth pursuing at all toward the understanding of competition with product variation.

We begin by sketching a rough taxonomy. This is by no means suggested as the final word. Undoubtedly there are overlaps in the classifications and omissions. The purpose is to isolate a few more categories than are usually found in one place in writings on oligopolistic competition and to suggest some possibilities for measurement and the construction of models.

The major categories to be considered are distribution and retailing; legal and institutional factors; production problems; advertising; public relations; gimmicks, games, and other price-related strategies; consumer-information activities; and finally, in a more basic category than the items noted above, models of consumer-information processes, both for the individual and for the mass.

Distribution and Retailing

Distribution and retailing systems vary heavily with the nature of the product. If the installation of a large turnkey system is the product being sold the prime manufacturer usually deals directly with the customer in a bargaining or bidding relationship. No separate distribution and retailing system is involved. For most mass market, small products both distributors and retailers exist. The distributors and retailers engage in locational competition. The location of sales outlets and of inventory points are of considerable importance. Many distributors and retailers are also concerned with activities such as grading and quality control. In many instances, service is a key factor. Competition between department stores and discount houses serves as a good example.

Service includes the politeness of personnel, packaging, credit facilities, home delivery, the size of the selection of close substitutes, or the number of different brands, varieties, or items carried. Convenience and promptness are also attributes of a good service system. In the sale of consumer durables, guarantees, warranties, and repair services are of considerable importance.

A good distribution and retailing system must not only take care of transportation, storage, handling, and delivery of goods, it must also accompany these processes with a dependable information network. The control of manufacturers over both the scope of the variables available to distributors and retailers and the speed of different reaction times varies considerably from industry to industry. These differences are undoubtedly critical for different types of competition. Interesting examples of the differences are supplied by the relationship between the large consumer durable manufacturers of items such as refrigerators or cooking ranges and their competition with a firm such as Sears. The automobile producers are another example. They work through only distributors where the distributor has a local franchise. Distribution of fruit is considerably different from any of these and in some instances the distributors may be far more powerful than the manufacturers.

To go from the general to the specific requires a great amount of empirical work. This does not mean that a general abstract theory cannot be constructed. It means, however, that it must lie at a level of abstraction that deals with strategies, flexibility of organization, viability, and other strategic concepts, which cannot be applied to a specific industry without a great amount of specification of the variables relevant to that industry.

Legal and Institutional Differentiation

Patents, copyrights, and customs play a considerable role in product differentiation. Patents and copyrights are discussed in great detail elsewhere (Bowman, 1973) and are not considered further here. It should be noted that custom is of considerable force in differentiation. Racial discrimination and the tendency of groups to "buy from their own" provide striking examples. Religious, national, and social observances provide considerable product variation. These are manifested in outright refusal to serve certain customers, religious dietary laws, and different standards of quality control enforced by custom as exemplified by English and French cuisine.

Production

There are several different ways in which production can be regarded as an extension of the marketing process. At the very least, it is worth considering three different uses of production as an extension of marketing structure: (1) new product production; (2) basic differentiation of an existing product; and (3) minor routine variation of a product.

A basic question when considering competition is whether people buy products or the services of products. If we adopt the latter viewpoint, the appropriate way of viewing a product is as a set of attributes. Viewed this way, items that may appear different are almost identical in some circumstances. Many a radically new product turns out to be in different competition with an item that may appear to be completely different. It is probably more accurate to say that consumers are interested in purchasing a product for its known attributes; hence a radically different new item must overcome consumer ignorance before it can succeed. The testing of new technology and the arranging for new production methods, coupled with the need to sell to a more or less uninformed market, means that new product introduction may easily have both a high level of variance in its outcome and take a considerable amount of time. If the weapon succeeds it may be highly successful. However, the individual firm has only partial control on the outcome and may often run a high risk.

A major product variation can be described loosely as one that requires nonroutine engineering to change the production line. Much of the discussion by Brems (1951) is devoted to this type of variation. Large teams of product development engineers may be maintained to work out important modifications and to establish the production processes needed to implement them. In some sense, basic modifications are not too far in character from product introductions, the major difference being that they generally take less time in development and involve less risk. The customers know more or less what to expect having already purchased the unmodified item. There is both an element of chance and a time lag involved in major modification. In particular this means that competitors cannot follow you immediately. Hence the addition to the degree of differentiation cannot be wiped out for at least a few months.

Routine product differentiation can be compared with using the different ingredients on the spice shelf to dress up the same piece of meat. If a competitor cuts a price or produces a special limited brand, a firm may wish to reply quickly and at the same time may not wish to precipitate a large-scale price war. One means of doing so is to put its own modification into the market. This can often be done in as little as six weeks. No basic change in production lines may be needed and the engineering for a whole set of minor modifications may have been carried out well in advance of the period for which they are needed.

Another important area of minor variation even more closely related to advertising is packaging. For some items such as chocolate bars or cigarettes, both quality and quantity variation may be of importance. Instead of varying the customer price of such an item it is possible to change its size or to modify the quality of the product.[1] In many instances in the production of supermarket shelf items, competition can be carried out through shading size. The sheer arithmetic involved in working out the per pound price of a jumbo package versus the regular package is one of the causes of the "truth in packaging" campaign. The design and production of attractive and in some cases especially useful packages and containers is an important part of product variation.

Advertising

In general, advertising is done by firms or other institutions that have as a goal the sale or other distribution of a product or service. Thus the goal of advertising is not to inform or instruct in a neutral manner but to aid in the fulfilling of the objectives of the advertiser.

Two distinctions are often made. They are advertising to inform and advertising to influence. The distinction is not always clear but nevertheless is useful up to a point. For example, in some cases, we may assume that a customer knows his tastes. He knows why he wants a particular soap or a special book but he does not know where to get it. An advertisement which says that the item is sold in a specific place and at a specific price may be regarded as providing the consumer with information. On the other hand, the consumer may have never heard of the product that a firm is selling. It may be new and initially the customer may not perceive a want for it. To employ a more technical phrase, the item may not originally be in the consumer's preference system. The role of advertising in this case is to introduce the item into the consumer's preference system in such a way that it is of sufficiently high value to move the customer to purchase it. In this case, it is easy to construct instances in which ac-

1. In England it has been estimated that a cut in cigarette length of one millimeter would add £30,000,000 to revenues. In Sunday *Times Weekly Review*, 8 June 1969, p. 45.

cording to the reader's own value system there will be "good and constructive advertising" and bad and "destructive advertising." The proponents and opponents of advertising mix technical considerations with value judgments combining them with few hard facts in such a way that it is generally easy to make out a case in either direction. In some instances, however, certain limited results can be obtained as we shall see in Chapter 11.

It is of interest to note that in one sense advertising can be regarded as an external economy to the consumer. After all, information and advice in many areas have a cost attached. The consumer interested in obtaining information much like the businessman interested in obtaining information does have the choice to pay experts. In certain purchases he does precisely that. For example, special medical services are often obtained as a result of paying a visit to one's general practitioner. The amount and value of information and communication are hard to quantify. Nevertheless, it is evident that these are becoming an increasingly important part of any modern economy.

The major weapons of advertising are television, radio, magazines, newspapers, handbills, and billboards. Advertising is usually executed by agencies. There is a middle man between the firm advertising and the media.

Although it is difficult to measure the amount and value of advertising expenditures, figures are often quoted. Critics claim that these figures are unduly high. Tax reasons help to boost the amount spent on advertising since, like any other factor of production, the expenses incurred can be deducted against cost. Leaving this aside, however, there are still several difficulties in interpreting advertising expenditure figures. The most important is when is an advertising expenditure in fact an advertising expenditure or when is it a form of discount or price rebate. In some instances a firm may give its distributors or retailers an advertising allowance that is not fully audited. Such an allowance is nothing less than a discount or price reduction done in such a manner that the list price of the item involved need not be moved.

In many instances it is difficult to distinguish between advertising and packaging or advertising and public relations. Given the vagueness of the boundaries there is a great amount of leeway in allocating advertising costs.

Public Relations

In some industries an important aspect of business strategy involves public relations. This is very often done in conjunction with advertising but is a somewhat different process. The phrase, it is not what you know but who you know that counts, is often used in a purely pejorative sense.

A consideration of information theory and communication systems will lead us to conclude quickly that good connections are a vital part of the global production process, which encompasses distribution and manufacturing.

As with advertising, undoubtedly there can be good and bad public relations. Many political lobbies and pressure groups leave much to be desired. The building of minor starlets into national figures strikes many of us as a dubious use of resources. It can be argued that newspaper and magazine articles are often used to purposely misrepresent an individual or organization. Fads are often associated with public relations and promotion routines. Trade associations use public relations to influence public opinion, bring pressure, and lobby at places of power.

Public relations are often employed in the preliminary steps to bargaining. In one sense they may be regarded as part of the preliminary process for obtaining a cooperative solution to a problem involving conflict among groups. Thus public relations as far as the large oligopolistic firm is concerned will be more closely associated with institutional advertising than with product information. The large firm will also probably be concerned with public relations concerning labor disputes and relations with the government. Certain professions such as the medical and legal professions, whose "ethical business practices" frown on price competition and advertising, are more concerned with their public image than they would otherwise be.

The symbiotic relationship between the producers and consumers of news makes it quite hard to evaluate the costs, expenditures, and services rendered in public relations activity. To a great extent the newspapers and weekly magazines need the "celebrities" as much as the celebrities need the newspapers and magazines. A crude measure, however, would be provided by the number of public relations firms, the number of press agents, and the public relations staffs of all corporations.

Gimmicks, Games, and Other Non-Price Alternatives

A complete array of minor weapons used at the manufacturing, distribution, and retailing levels exist that are not best described as either advertising or public relations. These involve games, give-a-way stamps, incentive schemes such as a free vacation trip, free samples, and bonus vouchers. All of these weapons may be used as alternatives to pricing changes and discounts. They appear to be reinvented every few decades by the current "marketing genius."

Another item frequently used and difficult to analyze without an adequate dynamic theory is the regular sale. What is the regular price of an item? Every year with clockwork regularity the price of an item such as sheets is lower at the annual January white sale than it is in early December. The regular sale enables firms to practice a form of price discrimination and service variation in such a manner that the signals among competitors are not confused. A competitor who cuts prices at any one of these institutionally accepted periods is not interpreted as making a declaration of hostilities, signaling the commencement of a price war. Another alternative to pricing in competition is under the table discounting. Discounts are used to provide the firms with far greater flexibility without disturbing the market climate than would be possible by the mere use of price.

Consumer Information Organizations and Strategies

Consumer expenditure on information and value influencing services is undoubtedly considerably less than the expenditures by firms and other institutions. Nevertheless, it is considerably higher than is often estimated. The role of brokers and expert evaluators provides a clue to these expenditures. Housing probably supplies the most important example.

At the level of mass information service the *Consumers' Report,* fair-trade organizations, and government information services for the consumer together with courses in subjects such as household economics are of some importance. It is difficult to impute the value of education in this particular area. Those who have lived in underdeveloped areas know full well that it requires a certain amount of education to be able to operate various consumer durables with ease and efficiency. Furthermore it may be argued that education provides an important antidote to the ills of misrepresentation and hard selling. It may be that the best protection against adulteration and lessening of quality of goods or services is an educated group of consumers who care.

A fundamental and difficult question that has no easy answer considering the implications of freedom of consumer choice is how much should fools be protected from their own folly? In one sense it is easy to say that everyone should be protected from gross and willful misrepresentation. However, in some instances this concept is hard to define. In another sense there are serious problems concerning who should be the stewards of public taste and private activity. Who should decide whether people can be permitted to expend a great amount of their earnings at Coney Island, other amusement parks, or at the tables at Las Vegas? When is information misinformation? What is the relationship among taste, good taste, freedom of choice, and the control of industry?

Advantages and Disadvantages of Advertising

In his study of the economic implications of advertising, Firestone (1967) offers the following among the reasons put forward for the use of advertising: it may lower the costs of production by enlarging the market and enabling the firm to make use of mass production methods; it may aid in improving quality and in simplifying products; it may aid in standardizing and stabilizing production; it may improve the climate for innovation and aid in increasing the variety of goods marketed. At the level of distribution and retailing, advertising may reduce the amount of retail sales effort to make a sale. It may replace less economic by more economic sales techniques; for instance, advertising might be used as a replacement for direct door-to-door selling.

It can also be argued that advertising serves as an aid to the education of the general public. Furthermore it subsidizes the mass communication media which in turn provides essential communication services to the consumer. Advertising may add to the consumer's stock of knowledge and increase the consumer's capacity to discriminate.

Other reasons advanced in favor of advertising concern its indirect aspects. It can be argued that advertising helps to advance technology, raise real national income, and generally contribute to the improvement of the standard of living.

The opponents of advertising can muster an equally impressive case against it. It may be argued that advertising can easily misinform the consumer. It often contains deceptive messages. It concentrates on persuasion rather than information. It helps to create and sustain sensational journalism and sensational television. It offends aesthetic standards and spreads external diseconomies to the public in the form of billboards, noise, and other nuisances.

From the viewpoint of competition it may lead to particularly high monopolistic profits. It protects producers' market shares.It contributes to wasteful product differentiation where imperceptible differences are magnified out of all proportion. It contributes to price rigidity, to more concentrated market structures, and helps to restrain more constructive aspects of competition.

Advertising can be said to involve a waste of resources, to contribute to cyclical variations, raise distribution costs, encourage the purchases of additional trappings of little value, such as some specialized packaging, and can even adversely affect production costs and quality if too great an emphasis is laid on irrelevant attributes of the commodity.

The array of reasons on both sides is formidable and there is no simple answer to the good or the evils of advertising. What may be true for advertising by soap companies may easily be false by shoe companies. The argument over the virtues of advertising depends very heavily upon facts and the amount of existing facts is scarce. Furthermore, the arguments depend very heavily on the explicit statement of value systems and a normative attitude toward advertising. In general, the value biases of the proponents are clear; they have something to sell. The value biases of the opponents are harder to make explicit; they include overly rigid utilitarians, and those with doctrinaire, religious, and political viewpoints. Among the milder proponents and opponents are also those who are willing to observe that there are many open unsettled questions that need both better analysis and a more explicit statement of the role of values.

In Chapters 4–5 the concept of solution was stressed. In the argument about advertising this appears again. Basically a solution may be regarded as a normative recommendation or a behavioristic observation. If we consider solution from a normative point of view, then we must ask whether we wish to solve an economic problem from the viewpoint of efficiency, equity, control of industry, and so forth. These different alternatives may in some cases be mutually exclusive. In some situations, loss of efficiency may be a reasonable price to pay for gain in freedom in choice.

Here we are interested in advertising as one of the weapons in the arsenal of the competitor. We do not propose to deal with the welfare implications of advertising in any further detail than to raise considerably more questions than we answer. The need for research on the actual effects of advertising is enormous. In this chapter and in Chapter 11 we raise some of the analytical problems that must be solved before we can fully appreciate the effectiveness of the weaponry of advertising, let alone evaluate general questions concerning its worth.

Theories and Measures

How is the consumer influenced by advertising or by changes in product? What do we know about reactions to quality? There are currently several theories, some of which have been formulated mathematically, picturing the spread of information through a community by diffusion or contagion processes. The mythology of marketing is replete with stories, conjectures, and theories concerning the faithful customer and the satisfied homemaker. Various conjectures have been advanced concerning the value of satisfaction with a previously used product as being critical in determining the purchase of a new item. For example, if considerable satisfaction was derived from owning a General Electric refrigerator, then upon next purchase the customer may be expected to be biased toward that particular brand. If, on the other hand, the customer was dissatisfied, one would expect a brand switch.

One hears of the so-called faithful customer, one who is not influenced by advertising or extra marketing effort, yet at the same time there is a mythology concerning the effectiveness of "nail-and-switch" techniques. This is the technique whereby some particular apparent bargain is offered to the customer who comes into the store to find that there is only the display model that has been nailed to the floor and the salesman then proceeds to switch the customer to another item. There is a rich body of folklore on the effectiveness of loss leaders and the effectiveness of displays to capitalize upon impulse buying. Some interesting work has been done and relatively careful analysis carried out on the importance of carrying a large variety of inventory when handling fashion goods (Murray and Silver, 1966; Liff, 1969). Studies of consumer behavior are far more systematic now than even ten years ago (Bettman, 1971). However, these advances are still in a very early stage.

What then are the measurements and what are the theories? The easy way out of our predicament is to say in an authoritative manner that all of the problems are qualitative and intuitive and then pass on to something else. The other alternative is to realize that the problems of measurement whether we are dealing with product differentiation, advertising, public relations, or other information processes are difficult in the extreme. The second alternative approach calls for us to begin at an extremely basic level. What are the dimensions of product differentiation? What are the dimensions of advertising? Such a question is asked at a level preliminary to that of measurement.

It is easy to see upon a little reflection that even at this level difficulties are encountered that cannot yet be overcome. How do we measure information? What are the units? How can we say that one message contains more information than another? Can any meaning be attached to saying that some message is twice as effective or contains as much information as another?

Suppose we were able to answer questions concerning dimensions and units of measure. Suppose measures existed. With these alone we would still be devoid of theories and even with theories we might be devoid of methods of validation and application. It is easy for any fool to sit down and say that selling expenditures may influence the demand curve. The important question is, how and what does the shift mean? One could start to theorize in an if-then manner, arguing that, if the shift is of such and such a shape, then its implications will be so and so. This, at least, produces some nonvacuous results and places us in the position to explore whether the phenomena of economic life conform to the model. If they do, then we would have linked some first steps at theorizing with a part of economic reality.

Work in marketing (Montgomery, 1971), the approach of behavioral theorists toward the acts of the consumer, the type of investigations by the Michigan Consumers Survey Group, and the mathematical models of operations research (Gupta and Krishnan, 1967; Jones, 1970; Aaker, 1971; Kinberg, Rao, and Shakun, 1974), suggest a completely new approach that offers the possibility of linking the many strategic variables of the firm to the behavior and reactions of the consumer.

Strategic Variables and Models of Competition

In this chapter we have briefly reviewed a large battery of different weapons available to the individual competitor. Depending on the specifics of the product, service, or system being sold, one, another, or some combination of these variables may far outweigh the more traditional variables of price and quantity. For example, in some government contracts quality and punctuality in delivery may be far more critical than price.

The methods of economics and microeconomics in particular teach us to deal with the "economic problem." This problem is usually posed in some form of cost minimization or profit maximization. However we pose it we need not necessarily talk about price or quantity as the major variables involved.

In the course of the years a certain amount of evidence has accumulated concerning the shape of cost curves or surfaces and the shape of demand with respect to price, with all other things being equal. If quantity and price are the relevant variables, we may link these variables to an economic model such that as they are varied we have some idea about the sensitivity of costs and profits. This is not the case in general for the other variables noted in this chapter. In abstracto, however, the problem is the same. The individual firm wishes to minimize costs or maximize profits and it has under its control a set of variables that need not be specified. There is a functional relationship between these variables and profit and costs. The firm's problem is to determine optimum levels of the values of these variables.

The economist's problem, if he wishes to extend his theory of competition, is to know not only how to prescribe the economic problem but how to go about selecting the most relevant variables for the market under study. Important variables for the automobile industry are not those for the tobacco industry or for public utilities, ethical drugs, or the supply of computer systems. Even knowing how to select the correct variables does not solve problems of measurement or the sensitivity of the reaction of the system as these variables are changed.

Our approach must be at two levels. The first is the abstract recognition that the economic problem is basically the same regardless of the names of the variables. To a great extent if we are talking at a level of generalities the mathematical models are identical regardless of whether the variables are called price, product variation, advertising, and so forth. However, given that we understand the mathematical formulation of the problem there remain two further steps to be taken. They are the identification of the relevant variables, the devising of appropriate measuring schemes for them, and the description of their relationship to goals and profits of the firm. Without the first step we would not have an adequate general conceptual understanding of the nature of the economic problem. Without the remaining steps we could scarcely advance to explanation of the economic phenomena that any good economic theorizing should be able to accomplish.

11 | Marketing Models in Noncooperative Equilibrium Analysis

UNDERLYING MUCH of oligopoly theory is the concept of noncooperative or quasicooperative behavior. When we analyze oligopolistic competition with product differentiation, advertising, and other variables present, the concept of a noncooperative equilibrium still holds and is of considerable use for analysis.

The general structure of the Prisoner's Dilemma game in which individual rationality leads to joint disaster was presented in Chapter 4. Price cutting strategies in markets for only mildly differentiated products also have this property. If one individual can cut price while the other or others do not, he gains considerably. If, however, they follow his lead all stand to lose. In some markets, such as cigarettes, a price strategy leads to a situation that is militarily similar to a nuclear bomb strategy. The major competitors each have the weapon. They can use it with relative accuracy and they know the damage it causes. If one uses it and the other does not retaliate, the user stands to gain considerably. If, however, there is retaliation, or if both use price strategy simultaneously, the result is disaster to all.

Perhaps one of the most important features of advertising, product variation, public relations, and new product introduction is that the variables are ill defined, ill controlled, have a somewhat unpredictable result, and invariably involve time lags. All of these features make them far more attractive as weapons. Once any of these variables becomes predictable, it is most likely that competition via that variable will cease. It will have become too dangerous to use except possibly in a moment of desperation.

Although the noncooperative equilibrium analysis is central to most attempts to construct a static theory of oligopoly we are well aware that eventually a dynamic theory is needed. In examining dynamics using games of economic survival (Shubik, 1959b) it becomes more apparent than with static models that the key aspects of understanding competition are the ability to describe and measure flexibility, viability, vulnerability, staying power, and the ability to maneuver. These are strategic terms. The specific variables influencing them vary from industry to industry and firm to firm.

Some economists have attempted to use measures of cross-elasticity to describe oligopolistic markets. Undoubtedly measures of cross-elasticity provide important information concerning one aspect of the interlinkage of the powers and fates of several firms. However, measures of cross-elasticity are grossly inadequate to offer anything more than a taxonomy of highly limited use. It may be that in the study of oligopoly the theorizing lies at a much more fundamental and abstract level than has usually been suggested. It lies at the level of thinking about strategies and viability in abstracto without naming the variables. On the other hand, the linkage between the theories and the reality lie at a far more concrete or institutional level than has usually been suggested. Thus, considerable detail is needed and an institutional understanding must be available to proceed from the abstract discussion of strategy space and flexibility to the concrete discussion on the nature of competition in a specific industry.

Marketing, Search, and the Serving of Customers

The subject of marketing involving distribution, retailing, sorting, grading, inspecting, packaging, warehousing, transportation, advertising, market surveying and so forth cannot be dealt with here. A few connections can be made, however, showing where this important subject connects with economic theory and more especially oligopoly theory.

Aggregation and Contingent Demand

An important topic that requires detailed microeconomic analysis concerns the segmentation of markets according to different groups of customers. Ordinary demand theory treats all customers as a single aggregate. A full understanding of demand in an oligopolistic market calls for a specification of how the population of customers is divided among the competing firms.

Possibly the earliest explicit mathematical model offered by an economist to investigate this problem was that of Hotelling (1929). He considered a set of customers spread out evenly along a line with a firm at either end. The delivered price to any customer is the price charged at the firm plus transportation costs, which are assumed to vary in proportion to distance. Variations and complications of this basic idea that considers all customers individually have been treated by Beckman (1968), Lovell (1970) and others.

If all customers and firms could be regarded, as a first approximation, as being located at a single point and if the firms were selling undifferentiated products, there is still a need to consider how the market splits when the firms have the power to name different prices but cannot individually satisfy all demand. A simple example somewhat elaborated from a previous publication (Shubik, 1959b) is given.

Suppose that an aggregate demand schedule of $q = 10 - p$ is given. However (as for example in the case of a newspaper) suppose that each customer wants only one unit and that their preferences are such that one customer is willing to pay 10 or less, another $10 - \epsilon$, a third $10 - 2\epsilon$ and so forth. Many different collections of individual customer profiles will aggregate to the same overall demand function. If a uniform price is always named by the suppliers and they have enough stock to serve all customers, the disaggregate information on the preferences and distribution of customers does not matter. If this is not so, this information is essential.

Figure 11.1 shows the structure of demand. Figure 11.2 shows the structure of contingent demand under two different hypotheses of market splits with different prices charged. These illustrations are given for duopoly but they can be easily generalized for more than two firms.

Suppose that the first firm charges a price p_1 where $p_1 < p_2$ and offers a quantity q_1^* for sale where q_1^* is less than $10 - p_1$. If all customers were informed we would expect the first firm to sell all of q_1^*. Given that this is the case what remains for the second firm? This depends on which customers are served by the first. If we were to assume that the customers are served randomly then the demand faced by the second firm will be q_2 as indicated in Figure 11.2(a); if the first firm were to service all the customers willing to pay a high price, the demand faced by the second firm would be q_2 as indicated in Figure 11.2(b).

In the first case the contingent demand faced by the second firm is given by DR'. We draw $R'F' = FR$ and connect D and R'. At any price above p_1^* a horizontal line across the triangle $R'DF'$ shows how much demand has been wiped out by the first firm. In the second case the contingent demand is formed by a parallel shift as the customers willing to pay higher prices are completely served. The contingent demand is ER'. In Chapters 5–7 we explicitly assumed that all customers had identical utility functions hence this extra problem in the determination of contingent demand did not matter.

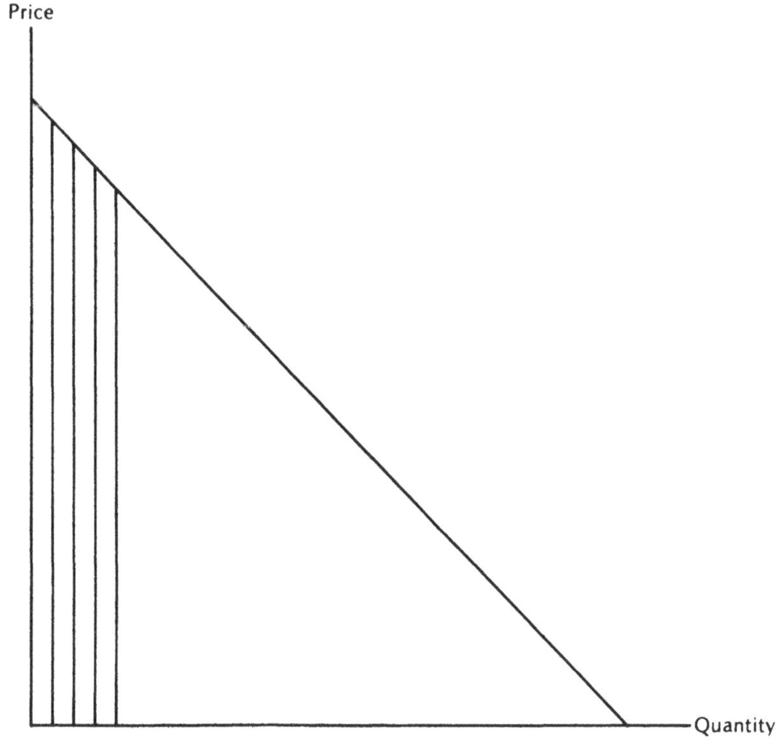

Figure 11.1. Disaggregate demand.

This somewhat esoteric construction which at first sight might be regarded as yet another complication and plaything for the duopoly theory writer, is a key link in connecting economic theory with market reality. How the customers select one firm over the other depends on transportation considerations, convenience in purchasing, and in many ways on the customer's state of information. For example, in the two methods for selecting contingent demand in Figure 11.2 we could interpret the difference in terms of information. In the first all customers are randomly equally informed of which firm has the lower price. In the second there is a search bias where those who are willing to pay more are more aware of where bargains are to be had. In actuality there is evidence that the latter hypothesis about the newspaper market would be false. The purpose of the illustrations however is to stress that this type of micro microeconomic model can handle this type of hypothesis whereas aggregate demand analysis avoids the key features of marketing.

The behavioral theorists of consumer behavior might argue that detailed rational search models are not the way to go. Regardless of this, the behavioral model of demand calls for the building up of the overall pat-

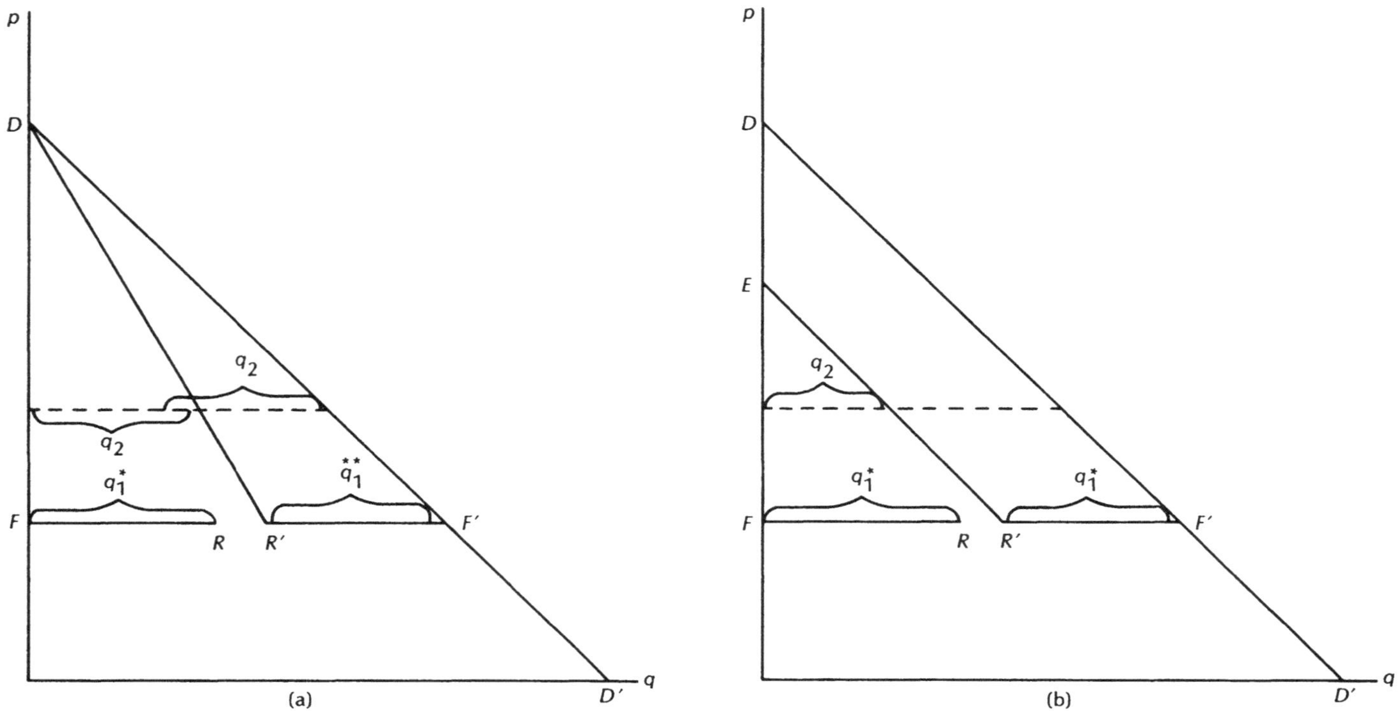

Figure 11.2. Market segmentation.

tern of demand from far more detailed features of the population than are utilized in much economic theory.

THE NEW INDUSTRIAL ORGANIZATION

Much of the basic mathematical modeling in marketing has come from operations research and management science. However two developments in economics have resulted in a new interest and different approaches to the study of industrial organization and competition and to the study of specific market mechanisms. They are the developments in the behavioral theory of the firm and mathematical models of incomplete information involving search, signaling, and models of bidding.

The developments in the behavioral theory of the firm can be seen in the works of Cyert and March (1955), Joskow (1973), Nelson and Winter (1974), and in an earlier attempt by Boulding (1962) to redirect some of economic analysis. This work in general is a mixture of microeconomic detail combined with technological, institutional, and behavioral factors usually neglected by the more traditional economic theorists.

The more formally mathematical work has emerged from the realization that the general equilibrium analysis (Walras, 1874; Arrow and Debreu, 1954; Debreu, 1959) of economic systems is essentially static and although it has much to say about a static price system it provides little help in explaining the evolution of prices. Furthermore the information conditions required for the existence of a general equilibrium are somewhat special as has been shown by Radner (1968) and others (Dubey and Shubik, 1977b).

When uncertainty and lack of symmetry in information conditions are introduced the models of economic behavior become extremely complex and the distinction between *so-called* rational economic models of behavior and the behavioral models is lessened. Rothschild (1973) has presented a summary of some of this recent work; the articles of McCall (1965; 1970) on the economics of information and job search and Spence (1974) on competitive and optimal responses to signals provide a sample of the type of problem considered.

A virtual subindustry has developed on the boundaries of economics and management science in the study of bidding and price formation mechanisms. Stark and Rothkopf (1979) provide a considerable bibliography and the works of L. Friedman (1956), Vickrey (1961), Greismer and Shubik (1963), Smith (1967), Wilson (1967), and Rothkopf (1969) serve to indicate the types of models and solutions under investigation.

INFORMATION AND BEHAVIOR

The models with imperfect information noted previously verge on the boundary between the study of stochastic processes and noncooperative games. There is a small amount of literature in which an explicit concern with the game theoretic aspects is evidenced. The work of Borch (1968) on insurance provides an example. A stimulating discursive article by Akerlof (1970) provides several examples in which the strategic importance of information is of significance.

Another duopoly model is presented here to provide an illustration of information conditions and to consider the sensitivity of competition to changes in information. The lack of information in this example is on the producer side of the market (Shubik, 1973b).

The Cournot Model with Different Information When we talk about competitors moving simultaneously as is usually done in describing Cournot duopoly we do not mean that they necessarily act at the same moment but that each must act without information concerning the behavior of the other. The two simple game tree diagrams shown in Figure 11.3 are representations of two games that are strategically equivalent. For ease in exposition suppose that two competitors must each select either a low or high level of production. P_1 stands for the first competitor and P_2 for the second. In Figure 11.3(a) the first competitor is assumed to actually set his production before the second. But the second is not informed of this, as is indicated by the information set[1] containing both vertices at which the second must select his production. In Figure 11.3(b) the order of the moves is reversed, but since the information conditions are unchanged, no strategic difference has been made.

There are two other possibilities that are not strategically equivalent. Those are shown in Figure 11.4. In Figure 11.4(a) the first competitor moves and the other competitor is informed prior to making his move. In Figure 11.4(b) the situation is reversed. These have been studied in Chapter 5 under the production leader solution.

Depending on technology and the formal and informal communication systems, there is the possibility of an information leak. If the individuals act independently sometimes one will move first and sometimes the other will move first. These possibilities are illustrated in Figure 11.5 where P_0, the first to move, may be regarded as "Nature" who selects from four alternatives with probabilities of p_1, p_2, p_3, and p_4 (where $\Sigma_{i=1}^{4} p_i = 1$). It can be seen that this game encompasses all of the others previously described as special cases. For example, $p_1 = 1$ is equivalent to the situation in Figure 11.4(a).

In Chapter 5 the usual noncooperative equilibrium with simultaneous

1. Each node in the game tree represents a point at which an actual player or Nature must make a choice. Although the omnipotent referee may know what choice point the player is at, if the player lacks information he may be unable to distinguish between several choice points. This lack of information is portrayed by enclosing these nodes in a curve to indicate that they belong to the same choice set.

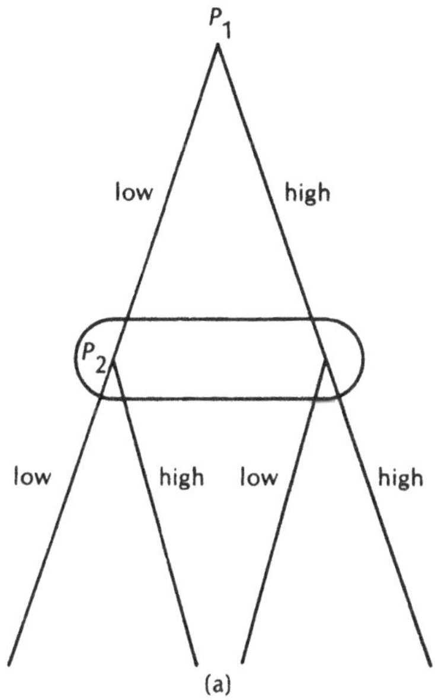

Figure 11.3. Duopoly with simultaneous information.

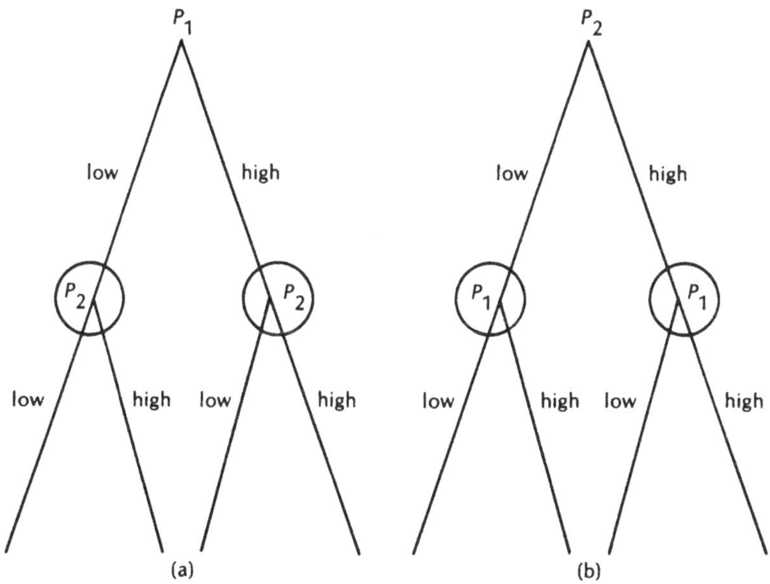

Figure 11.4. Duopoly with sequential information.

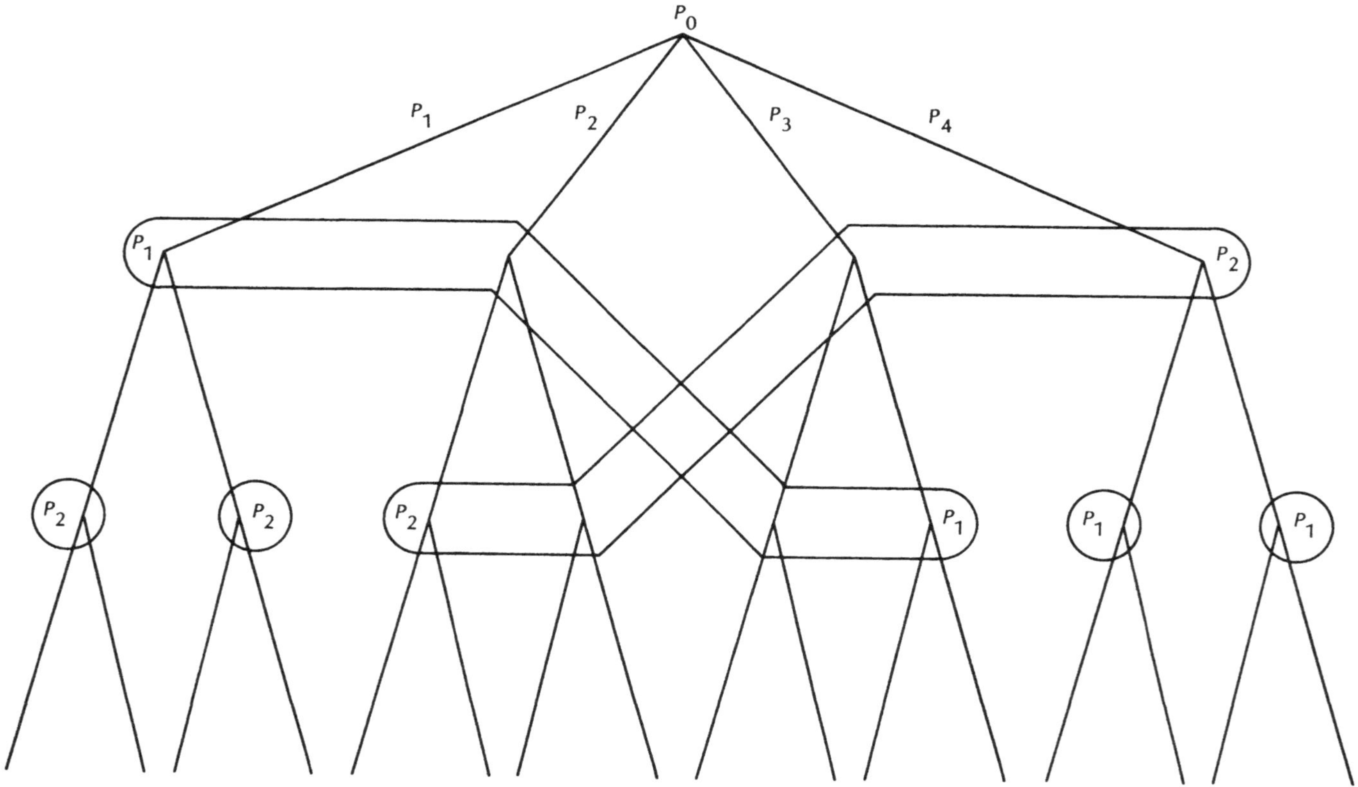

Figure 11.5. Duopoly with an information leak.

moves and the production leader models have been examined. We turn to the model illustrated in Figure 11.5. For ease of illustration suppose that all of the p_i are equal; hence any $p_i = 1/4$. There are four cases, but as is indicated by the information sets each player can only make two distinctions prior to moving. If he is told the move of the other he knows that he is in a situation akin to Figure 11.4(a) for the second player. If he is given no information, he knows only that he is in one of two simultaneous move games or that after he has selected his production the other player will be informed prior to making his selection.

For simplicity, set $c_1 = c_2 = 0$. An equilibrium strategy for P_1 is as follows:

$$\text{If } P_1 \text{ is informed of } q_2 \text{ select } \hat{q}_1 = \frac{\alpha - \beta q_2}{2\beta}.$$

$$\text{If}^2\ P_2 \text{ is not informed of } q_2 \text{ set } q_1 = \frac{5}{14}\frac{\alpha}{\beta}.$$

The symmetric equilibrium is given by

$$\hat{q}_1 = \hat{q}_2 = \frac{9\alpha}{28\beta} \quad \text{and} \quad q_1 = q_2 = \frac{5\alpha}{14\beta}, \tag{11.1}$$

with expected profits of

$$R_1 = R_2 = \left[\frac{1}{4}\left(\frac{9}{28}\right)^2 + \frac{1}{4}\frac{90}{(28)^2} + \frac{1}{2}\frac{80}{(28)^2}\right]\frac{\alpha^2}{\beta} = \frac{331\alpha^2}{3136\beta}. \tag{11.2}$$

The effect of symmetric uncertainty is to slightly increase the overall amount offered to the market. The full comparison among different solutions is given in Table 11.1. In Table 11.1, profits and production have been normalized so that they are in the range from 0 to 1. This makes it easier to discuss the magnitudes of the three effects of numbers, information, and communication.

Limiting ourselves to variation of information, but no communication of strategies or discussion among players the variation in individual player profits is from 0.25 to 0.5 and in joint profits from 0.75 to 0.888. In this instance the ability to transmit information is valuable to the player doing so.

An important distinction must be made between a general precommit-

2. P_1 tries to maximize

$$2(q_1[\alpha - \beta(q_1 + q_2)]) + q_1\left(\alpha - \beta\left[q_1\left(\frac{\alpha - \beta q_1}{2\beta}\right)\right]\right)$$

assuming symmetry we have $2\alpha - 4\beta q_1 - 2\beta q_1 + (\alpha/2) - \beta q_1 = 0$ hence $q_1 = (5/14)(\alpha/\beta)$.

Table 11.1 Solutions for duopoly with differing information conditions.

Solution	q_1	q_2	$q_1 + q_2$	P_1	P_2	$P_1 + P_2$
Collusion	—	—	1/2	—	—	1
Duopoly P_1 first, perfect equilibrium	1/2	1/4	3/4	1/2	1/4	.75
Duopoly simultaneous	1/3	1/3	2/3	.444	.444	.888
Duopoly random information	9/28,10/28	9/28,10/28	39/56	.413	.413	.816
Duopoly efficient	1/2	1/2	1	0	0	0

ment and information concerning a move. A general precommitment is a strategy or parts of a strategy with contingent clauses, such as "If you do x, I will do $f(x)$." Depending on one's model of human behavior we may attach various degrees of belief to whether the contingencies stated will actually be followed.[3] In contrast, information concerning a move is information concerning an act that has taken place. In these models the acts are the selection of levels of production. Communication of strategies is to information concerning moves as words are to deeds.

When we include communication with the sending of threats the advantage to having more information now goes to P_2 *if* his threats are believed. We note that once threats are included the players must be able to judge both the *ability* and the *intention* to carry them out. Thus in the case of duopoly, costs and capacities must be known.

Advertising and Marketing

Quantitative Models of Advertising and Marketing

There has been a major upswing of interest in the use of mathematical models in marketing since the late 1950s. The impetus comes primarily from management science and operations research with some interest from economists (Oxenfeldt, 1963) and the more recent work of mathematical economists. The general message in much of this work is the necessity to descend to the levels of micro-microeconomics when trying to understand a specific market. Put more simply those features considered merely frictions that do not matter in the long run by many economists have now been recognized as important. Abstract analysis alone is no substitute for knowledge of the institutional and technological facts of the business being analyzed. The needed facts are considerably more than most microeconomic theory prior to the 1960s has taken into account.

In Bass et al. (1961) mathematical models are constructed to study brand preferences, variety in retailing, spatial allocation of selling expenses, the relationship between optimal advertising and optimal quality. A section of this book is devoted to game theory models of advertising expenditures, promotional competition, and a model for budgeting advertising. The simple advertising model presented below is in the spirit of the work noted here.

Montgomery and Urban (1969) provide a management science view of advertising with a discussion of advertising goals and a consideration of factors such as exposure, awareness, and attitudes. Stochastic models of consumer response are also considered. Selling efforts, pricing, and new

3. In the construction of game models we may decide to model verbal statements as actual moves in the game, in which case the degree of belief must be specified in the rules. Frequently words and the communication system are left out of the game.

product introduction are discussed. Other literature (Dorfman and Steiner, 1957; Nerlove and Arrow, 1962; Gabor and Granger, 1966; Leitch, 1974) considers linking market strategy with production scheduling and problems of marketing mix (i.e., tradeoffs between quality, price, advertising, and salesforce).

A Specific Model of Advertising

In this section a specific and somewhat ad hoc modification of the model presented in Chapters 6 and 7 is made so that advertising can be taken into account. Demand which was presented in Chapter 7 as

$$d_i = \frac{1}{n}(\alpha - \beta[p_i + \gamma(p_i - \bar{p})]) \tag{11.3}$$

must now be presented as a function of $2n$ variables; the prices and advertising expenditures. This is expressed as

$$d_i(p, a) = \frac{1}{n}(\alpha - \beta[p_i + \gamma(p_i - \bar{p})])[(1 + \eta)\sqrt{\Sigma a_j}]f_i(\theta, a) \tag{11.4}$$

where $p = (p_1, p_2, \ldots, p_n)$, $a = (a_1, a_2, \ldots, a_n)$, and

$$f_i(\theta, a) = \begin{cases} \theta + (1 - \theta)\dfrac{na_i}{\Sigma a_j} & \text{for } \Sigma a_j > 0, \\ 1 & \text{for } \Sigma a_j = 0. \end{cases} \tag{11.5}$$

The d_i are constrained to be nonnegative.

Institutional Advertising The effect of advertising is divided into two parts: institutional or that which might aid the industry as a whole, and individual or that which benefits one firm by taking customers away from the others.

The term in Eq. (11.4) which is given by

$$(1 + \eta)\sqrt{\Sigma a_j} \tag{11.6}$$

controls the overall or institutional effect of advertising. The parameter η controls the effectiveness of the overall impact on industry demand. If η equals zero, advertising has no effect on the overall demand in this industry. A square root is introduced to act on the sum of advertising expenditures to produce the effect of diminishing returns.

Figures 11.6 and 11.7 show two means of introducing the overall industry effect of advertising. In the model constructed here, the effect as indicated in Figure 11.7 has been used.

For the partial shift effect, the terms involving advertising need to be applied only to the constant α in the demand function. Since a case on empirical grounds can be made for either way of treating advertising, the second was chosen for ease of computation. As can be seen in the calcu-

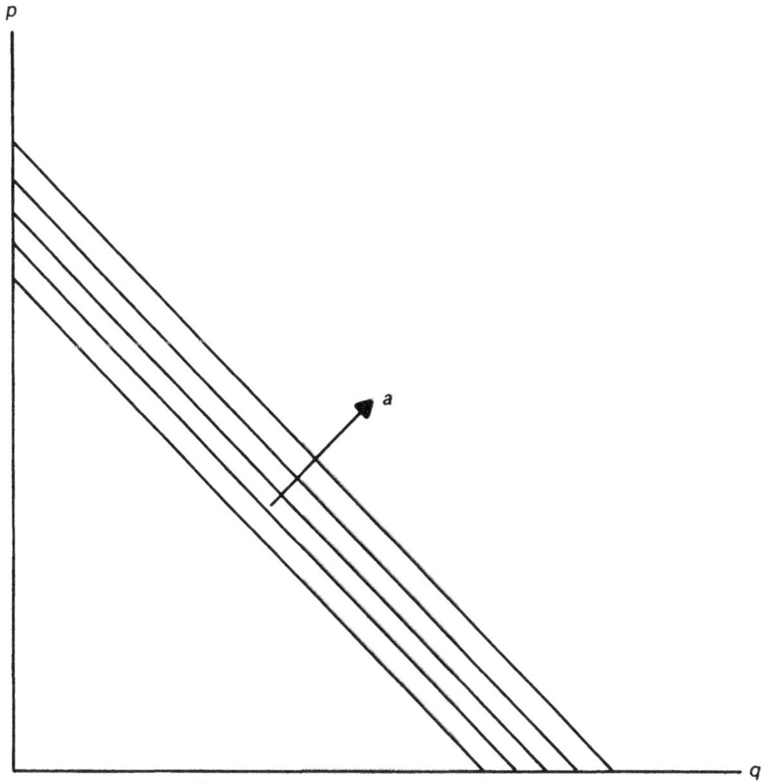

Figure 11.6. Institutional advertising.

lation of the solutions given below the optimization conditions give equations with a triangular form as the advertising effect enters in a multiplicative manner. Not only does this mean that the equations are easier to solve but they imply that the optimal price does not depend upon the level of advertising although the optimal level of advertising in this case depends upon the selection of price. This feature offers better experimental control for studying behavior in the setting of advertising budgets when the model presented here is used as an oligopoly game.

The selection of the particular form given in Eq. (11.6) is undoubtedly ad hoc, yet since the empirical basis for modeling institutional or industry-as-a-whole advertising is not well developed a reasonable and easily analyzed form was selected.

Individual Differentiating Advertising The term given in Eq. (11.5) describes the competitive effect of advertising. The parameter θ, between zero and one, controls the competitive aspect of advertising. If this parameter is zero, the apportionment of the market depends completely

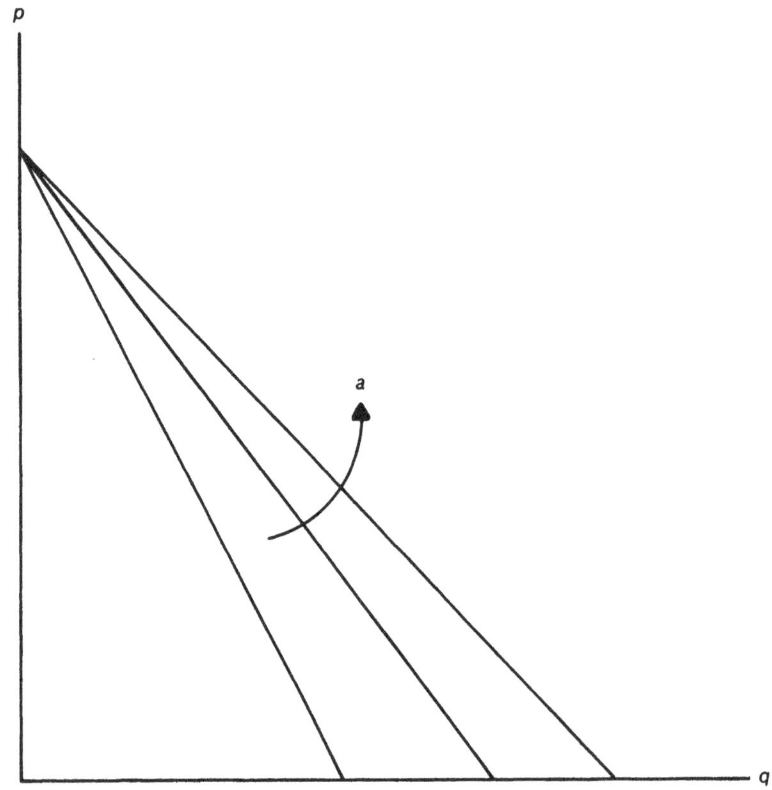

Figure 11.7. Competitive advertising.

on advertising; if it is one, advertising has no competitive component. This particular functional form for advertising is well known and reflects two features of the folklore of advertising. In particular, θ may be considered the percentage of customers who are not influenced by advertising. The remaining $(1 - \theta)$ customers are distributed among the firms in proportion to each firm's share of total advertising.

SOLUTIONS

In this section the values of advertising expenditures are determined for the n-firm symmetric oligopolistic market. The joint maximum, price noncooperative equilibrium, and beat-the-average solutions are investigated.

Joint Maximization Suppose that each firm takes a highly cooperative attitude toward all the others. Consider that each firm acts to maximize the sum of the profits of all firms. By symmetry, we may assume all prices and advertising budgets equal and maximize the profit of one firm.

In a symmetric solution, profit for each firm is given by

$$P = \frac{1}{n}(p - c)(\alpha - \beta p)(1 + \eta\sqrt{na}) - a - k \tag{11.7}$$

(where a is the individual expenditure of advertising k, the fixed costs), since by symmetry the competitive advertising factor equals 1, and $p_i = p$ for all i.

Since a necessary condition for maximization is $\partial P/\partial p = \partial P/\partial a = 0$,

$$\frac{\partial P}{\partial p} = \alpha - 2\beta p + \beta c = 0 \tag{11.8}$$

and

$$\frac{\partial P}{\partial a} = (p - c)(\alpha - \beta p)\frac{\eta}{2\sqrt{na}} - 1 = 0. \tag{11.9}$$

Solving we get

$$p = \frac{\dfrac{\alpha}{\beta} + c}{2} \tag{11.10}$$

as before, and

$$a = \frac{1}{n}\left(\frac{\eta(\alpha - \beta c)^2}{8\beta}\right)^2. \tag{11.11}$$

Noncooperative Equilibrium (Price and Advertising) If advertising is held constant, the remaining price game analysis falls into two categories that can be described as the pure strategy equilibrium (Bertrand or Chamberlin case) or the "cycling" equilibrium (Edgeworth case). The shape of the contingent revenue functions will determine which of the two types of solution exists. The analysis is limited here to the Chamberlin case, although as noted in Chapter 8, inventory and stockout cycles are probably the rule rather than the exception.

The payoff function of the individual firm can be expressed as

$$P_i = \frac{1}{n} G_i(p)H_i(a) - a - k, \tag{11.12}$$

where

$$G_i = (p_i - c)\left[\alpha - \beta\left(1 + \frac{n-1}{n}\right)\eta p_i + \frac{\beta\gamma}{n}\sum_{j\neq i} p_i\right], \tag{11.13}$$

and

$$H_i = \left[\theta + (1 - \theta)\frac{na_i}{\Sigma a_j}\right](1 + \eta\sqrt{\Sigma a_j}). \tag{11.14}$$

For a maximum of P_i with respect to p_i the necessary condition is

$$\frac{\partial P_i}{\partial p_i} = \frac{1}{n} H_i \left(\frac{\partial G_i}{\partial p_i}\right) = 0, \tag{11.15}$$

which implies

$$\frac{\partial G_i}{\partial p_i} = 0. \tag{11.16}$$

Since the G's are functions only of the prices, we may search for an equilibrium-point solution of a price subgame independently of the whole game. This is the same solution as given in Chapter 7:

$$p = \frac{\frac{\alpha}{\beta} + c\left(1 + \frac{n-1}{n}\gamma\right)}{2 + \frac{n-1}{n}\gamma}. \tag{11.17}$$

The situation, with respect to advertising, is somewhat more involved. A value for advertising by the firm is calculated. We assert, deferring the proof, that this value of advertising together with the price gives a symmetric equilibrium point for the game. The question of whether this game has asymmetric or mixed strategy equilibria is still open. We conjecture that it has none.

Again, for nonzero advertising, the necessary condition for an equilibrium is

$$\frac{\partial P_i}{\partial a_i} = \frac{G_i}{n}\frac{\partial H_i}{\partial a_i} - 1 = 0. \tag{11.18}$$

Let us call $G_i(p_{nc}) = G_{nc}$, where p_{nc} indicates the value of p obtained from Eq. (11.18), hence we have

$$\begin{aligned} \frac{n}{G_{nc}} &= \frac{\partial H_i}{\partial a_i} \\ &= \frac{\eta}{2}(\Sigma a_i)^{-1/2}\left[\theta + (1-\theta)\frac{na_i}{\Sigma a_j}\right] \\ &\quad + [1 + (\Sigma a_j)^{1/2}](1-\theta)\eta\frac{(\Sigma a_j - a_i)}{(\Sigma a_j)^2}. \end{aligned} \tag{11.19}$$

Searching only for symmetric solutions to this set of simultaneous nonlinear equations, let $a_i = a$ for all i. Thus,

$$\frac{\eta}{2}(na)^{-1/2} + [1 + \eta(na)^{1/2}](1-\theta)n\frac{(n-1)}{(na)^2} = \frac{n}{G_{nc}}. \tag{11.20}$$

Hence, since $na = 0$ is not a root of this equation,

$$\frac{n}{G_{nc}}(na) - \eta\left[(n-1)(1-\theta) + \frac{1}{2}\right](na)^{1/2} - (n-1)(1-\theta) = 0. \quad (11.21)$$

This quadratic equation in $(na)^{1/2}$ is solved by the standard formula and can be seen by the rule of signs to have one positive root. Hence,

$$a = \frac{G_{nc}^2}{4n^3}\left(\eta\left[(n-1)(1-\theta) + \frac{1}{2}\right] + \sqrt{\eta^2\left[(n-1)(1-\theta) + \frac{1}{2}\right]^2 + 4\frac{\eta(n-1)(1-\theta)}{G_{nc}}}\right)^2 \quad (11.22)$$

where

$$G_{nc} = \beta\left(1 + \frac{n-1}{n}\gamma\right)\left(\frac{\frac{\alpha}{\beta} - c}{2 + (n-1)/n}\right)^2.$$

We omit a proof that the strategy pair a_{nc}, p_{nc} constitute an equilibrium solution.

Beat-the-Average Solution The beat-the-average solution has been discussed previously. An equilibrium solution of the transformed game with payoffs

$$\prod_i = P_i - \frac{1}{n-1}\sum_{j\neq i} P_j \quad (11.23)$$

is sought.

There are two motivations for considering that a solution of this transformed game is of predictive value. First, in the game playing experimental setting, it has been observed that players may interpret their task as that of out-doing the other players currently in the game. This is a highly competitive style of play and game players especially in the absence of monetary reward seem to seek relative rather than absolute gain. Second, if one seeks to apply the Nash (1953) bargaining theory to an n player oligopoly, it turns out that the beat-the-average solution is an optimal threat point in the game with sidepayments. Hence, we might expect that this solution is of use in examining the threat structure in cartelized markets.

We proceed as in the other cases and solve for first-order conditions. However, we use the payoff transformation given in Eq. (11.23), and write P_i for π_i. Again, we search only for symmetric solutions and the special form enables us to solve for p independently, obtaining the same value as in Chapter 7:

$$p_{ba} = \frac{\frac{\alpha}{\beta} + (1 + \gamma)c}{2 + \gamma}. \tag{11.24}$$

To solve for advertising we have

$$\frac{\partial P_i}{\partial a_i} = \frac{G_i}{n}\frac{\partial H_i}{\partial a_i} - 1 - \frac{1}{n(n-1)}\sum_{j \neq i} G_j \frac{\partial H_i}{\partial a_i} = 0. \tag{11.25}$$

Since, by the symmetry assumption $G_i = G_j = G_{ba}$ for all i and j, we get

$$\begin{aligned}\frac{\partial H_i}{\partial a_i} - \frac{1}{n-1}\sum_{j\neq i}\frac{\partial H_i}{\partial a_i} = \frac{n}{G_{ba}} \\ = \frac{n}{2}(\Sigma a_j)^{-1/2}\left[\theta + (1-\theta)\frac{na_i}{\Sigma a_j}\right] \\ + [1 + \eta(\Sigma a_j)^{1/2}]\frac{(\Sigma a_j - a_i)(1-\theta)n}{(\Sigma a_j)^2} \\ - \frac{1}{n-1}\sum_{j\neq i}\left[\frac{\eta}{2}(\Sigma a_k)^{-1/2}\left(\theta + (1-\theta)\frac{na_j}{\Sigma a_k}\right)\right. \\ \left. + (1 + \eta(\Sigma a_k)^{1/2})\frac{n(1-\theta)(a_j)}{(\Sigma a_k)^2}\right].\end{aligned} \tag{11.26}$$

And since by symmetry $a_i = a_j = a_{ba}$, we have

$$[1 + \eta(na_{ba})^{1/2}]n(1-\theta)\frac{1}{(na_{ba})} = \frac{n}{G_{na}}.$$

Solving the quadratic equation in $(na_{ba})^{1/2}$ we get

$$a_{ba} = \frac{G_{ba}^2}{4n}\left[\eta(1-\theta) + \sqrt{\eta^2(1-\theta)^2 + 4\frac{(1-\theta)}{G_{ba}}}\right]^2, \tag{11.27}$$

where

$$G_{ba} = \beta(1+\gamma)\frac{(\alpha/\beta - c)^2}{2+\gamma}.$$

Some Conceptual Problems In Chapters 5, 6, and 7, the competitive or efficient solution was investigated. Formally we could set up the conditions[4] here and obtain $p = c$. However, in extending this solution to advertising we would have to make the assumption that the firms not only know the current preferences of the individuals, but perceive how their preferences change and could, if they wished, determine how much advertising "is good for the customer." This would call for a basic modifica-

4. The limiting case of the beat-the-average solution as $n \to \infty$ is also $p = c$.

tion of the underlying consumer preferences given in Eq. 6.1, to include the change in preferences as a function of advertising. At this point, even the most unreconstructed neoclassical economist must bow to the social psychologist, operations researcher, and marketing professional. Further, ad hoc modeling without much in the way of theory, empirical data, or experimental information does not appear to be too fruitful.

With advertising the breakeven solution is not uniquely defined. There will be a continuum of breakeven solutions. In the solutions presented here fixed costs k were noted in the payoff functions. Although they play no role in the joint maximum, beat-the-average, and price parameter solutions, they do influence the noncooperative equilibrium if they are sufficiently high. This is left as an exercise for the reader.

Product Differentiation and Variation

The parameter γ in Eq. (7.7) controls the level of product differentiation. We could treat it as a variable rather than a constant. Thus, if we let z_i = the amount spent by the ith firm on product variation, $\gamma_i = \gamma_i(z_1, z_2, \ldots, z_n)$. Hence, Eq. (7.7) becomes

$$P_i = \frac{1}{n}(p_1 - c)(\alpha - \beta[p_i + \gamma_i(z_1, z_2, \ldots, z_n)(p_i - \bar{p}]) - z_i. \quad (11.28)$$

This is formally like the advertising models and we could solve for a noncooperative equilibrium with p_i and z_i as the strategic variables. Intuitively however, product variation appears to be different from advertising, although there may be borderline cases. In particular, the nature of the investment, the length of time involved, and the uncertainty associated with the expected results are considerably different for basic research and development of new products, cosmetic changes of existing products, packaging, and advertising.

Without being explicit about the dynamics and the uncertainty associated with product variation and advertising, it is scarcely possible to make a satisfactory operational distinction in models such as those shown in Eqs. (11.5) and (11.6). This point was forcibly illustrated in a pilot gaming experiment. Graduate students in economics were asked to play in a duopolistic market with structure as in Eq. (11.4); however, a scenario was provided for half of the students describing the advertising terms as advertising, and for the other half describing them as product development and variation. Several of the students in the latter group complained about the lack of realism of the model; one even observed that the variable seemed to act more like advertising than product variation, but beyond that behavior was not significantly different.

It must be noted that this difficulty in making a distinction between the two appears in accounting. Advertising is usually treated as an expense or cost, whereas research and development may be treated as an investment or an expense, although laws, customs, and accounting practices differ in different countries on this point.

An immediate observation concerning product differentiation and antitrust can be made. The Robinson Patman Act provides an incentive for large firms who sell in many markets to resort to cosmetic changes of product to enable them to charge different prices for virtually the same good in different markets.

Appendixes
Bibliography
Index

Appendix A | Accounting, Economic Theory, and Gaming

ONE OF THE IMPORTANT BY-PRODUCTS of using a business game linked to a formal economic model is that the players are able to see the connection between the economic structure of the market, balance sheets, profit and loss statements, and competitive market information in a way that is far more realistic and relevant than learning either the economic theory or the accounting separately.[1] Accounting serves many different purposes and has conventions that from one viewpoint seem inadequate or irrational but from another serve the decision or measurement purpose at hand. The linkage between statics and dynamics in economic theorizing has always been tenuous at best. Many of the problems in accounting illustrate the difficulties in successfully marrying static and dynamic economic theory. Accounting is still so much of an art; it is well known in large bureaucracies and financial institutions, such as brokerage houses, that "creative accounting" can project a very different picture of the same facts to the naive.

It is of paramount importance that the microeconomic theorist see clearly the relationship between a model of the market and the accounts and other reported information on the firms. In a market structure as simple as that presented here, it is quite evident that we can scarcely go into the arcane arts of accounting. Nevertheless, the simple economic structure can be tied in a straightforward manner to the regular features of

1. There are many useful basic texts. A brief review is given by Mauriello (1971).

a profit and loss statement and a balance sheet. Furthermore, as we have seen from our discussion of demand and capacity restrictions, we can employ some market statistics that are unavailable in the real world but that nevertheless illustrate important problems in market information gathering and in the measurement of competition. In particular, we can report lost sales.

INDUSTRY AND INDIVIDUAL STATISTICS FOR THE MARKET GAME

The following presents a copy of the type of output received by one of the competing firms during each quarter of the play in a game having the same structure as the oligopoly models in Chapters 5 through 8.

GAME 1A, 2 FIRMS

COPY FOR COMPANY 2	COMPANY 1	COMPANY 2	AVERAGE
Price ($)	240.	240.	240.
Sales (Units)	200,000.	200,000.	200,000.
Advertising ($)	4,000,000.	9,000,000.	6,500,000.
Production (Units)	250,000.	200,000.	225,000.
Inventories (Units)	300,000.	250,000.	225,000.
Net Profits ($)	5,249,996.	2,374,996.	3,812,496.
Dividends ($)	0	0	0
Investment ($)	0	0	0
Rate of Interest	3%	3%	3%
Present Value of Dividends Paid to Date ($000)	0	0	0
Lost Sales (Units)	0	0	0

In using most business games it is customary to give each of the firms one or two quarters' worth of information prior to asking the players to make decisions. In this manner, it is possible to make preliminary estimates concerning price sensitivity and general structure of the market. Furthermore, such a procedure can be justified in terms of conceiving of new management taking over the operation of an ongoing firm and having a statistical history of the firm with which to work.

Price is the price per unit of the product that each firm sells. Since the products are not identical, although one can define average price, one must be careful in interpreting it. It is an aggregate in the same way that any price index is an aggregate. Sales are the number of units a firm has sold over the period which in the case of the game is a quarter. Advertising is the amount of money spent on advertising during the quarter. There is, of course, no guarantee that advertising is immediately useful. One of the problems faced by any firm is determining the sensitivity of its market to advertising and the time effect of advertising. In some cases, it may take many quarters of advertising before any impact is felt.

Production is specified in units; these may be individual items such as automobiles or cases, cartons, gallons for small products such as cigarettes or beer. There is an upper bound to the amount that an individual firm can produce in any time period. This amount is determined by the capacity of the firm. It is assumed that the quarterly production is produced smoothly over the whole of three months.

The figure for inventories represents the inventory available to the firm at the end of the quarter for which it has received a report. It is important to note the relationship among production, inventory, and sales. It is assumed that only half of this quarter's current production will be available for sale during any particular period. Since production and distribution are assumed to take time, customers may be lost as a result of delays if they cannot be supplied from immediately available inventories. This means that we can write equations for sales in period t and inventories available at the end of period t or at the start of period $t + 1$ as follows:

$$S_{i,t} = \max[d_{i,t}, I_{i,t-1} + \tfrac{1}{2} q_{i,t}],$$
$$I_{i,t} = I_{i,t-1} + q_{i,t} - S_{i,t},$$

where $S_{i,t}$ represents sales of company i at period t; $d_{i,t}$ is the demand faced by company i at period t; $I_{i,t}$ is the inventory of company i at period t; $q_{i,t}$ is the production of company i at period t. The calculation of net profits will be presented below during discussion of the profit and loss statement. In this particular example, dividend payments were zero. However, dividends may be paid by the players and will be recorded in this particular position.

The next piece of information is the amount of investment. In games in which the individuals are called upon to make investments, the price of an added unit of capacity is announced and the players are informed of the existing time lag which determines the waiting period between ordering new capacity and having the new capacity in production.

The rate of interest is the quarterly rate of interest that specifies the value of money in alternative employment. If we were to assume that a perfect capital market existed, earnings on dividends would also be at the same rate as bank-rate borrowing. Usually this is not the case. In general, an individual can earn less when he invests his dividends than the company pays for money borrowed from the bank. In some cases, the game may be played with two rates of interest, one rate representing earnings on dividends, and the other rate representing the price of money from the bank.

The next two items may seem somewhat puzzling. The first is the present value of dividends paid to date. In our exploration of the behavior of the firm we have added financial considerations. Although it is not the only goal that the firm may pursue, one reasonable economic goal is to maximize the sum of present value of dividends paid and its current net worth. One way of looking at this is as follows: suppose that you are the sole proprietor of a corporation. Any dividends that you pay out are paid to you. You may spend them in any way you wish, or in particular, you have the choice of investing them. The value of money is given by the rate of interest. Hence, you could invest all of your dividends at the going rate of interest. Thus, if we wished to tally your performance at any point, we would not only evaluate the net worth of your corporation but would also evaluate the current net worth of your dividends plus the interest they will have earned. Idle cash balances lying around a firm do not produce extra income, hence it will pay you as owner of the corporation to employ the money. Excluding important tax considerations,[2] you can at least obtain the going rate of interest by paying yourself the idle cash balances as dividends.

All firms require a certain amount of cash to run the daily business and to guard against short-term fluctuations, so not all cash is necessarily idle. Some must be kept for transactions and short-term protection.

If the bank rate is higher than the earning rate on dividends, a manager will have to think twice before paying out dividends if his short-term assets are sufficiently low that he may have to use a bank line of credit. In actuality a healthy firm can make a better return on money than can its smaller stockholders. This follows immediately from their differentiation of function, that is, a firm usually is better informed and has more investment opportunities than a small stockholder. The firm should contain professionals, whereas the small stockholder will neither be as competent nor as well aware or organized to take advantage of opportunities as is the corporate investor.

The last item of information is lost sales. This is a mystical concept. It provides us with an excellent example of the fallacy of composition. Industry lost sales, when considering the industry as a whole, can easily be far fewer than the sum of all lost sales. A simple example illustrates this point. Suppose ten firms are all selling the same item at the same price. Furthermore, suppose that after a brisk period of sales, they all happened to run out of inventory at the same time. For instance, suppose that the item is a magazine, which quickly sold out. It is possible that the supply was one unit less than the total demand. However, the extra customer could conceivably appear at the first point of supply, ask for a magazine, and be told that it is out of stock. He would then be counted as a lost sale. He could go to the next firm, do the same there, and so on through the whole ten points of supply. As a result, each firm would report a lost sale of one unit. However, the aggregate lost-sale to the industry is also one unit, and not ten units. The possibility of this phenomenon leads to a

2. In this model we do not account for individual income taxes but do include corporate income taxes. When an individual is in a tax bracket higher than that of a corporation in which he owns shares, he may prefer to see its profits used internally thereby building up its asset value rather than obtaining dividends that will be taxed heavily.

serious danger of cyclical fluctuations in ordering policy owing to gross overestimation of total lost sales.

In Chapter 5 we discussed the theory of contingent demand and rationing. The effects of lost sales or stockouts of one firm on the sales of another are complicated to estimate. Undoubtedly a more dynamic theory of demand than that presented in Chapter 5 is needed. Customers have memories and this period's stockout may influence next period's demand as well as this period's demand. This is the type of problem that stresses the need for a marriage of economic and marketing theory with a larger behavioral understanding of the acts of the consumer through time.

The Profit and Loss Statement

The following presents the profit and loss statements and the balance sheet. Net sales billed is the amount of money obtained from the sale of the product. In other words, it is quantity sold times price.

QUARTER 2, GAME 1A, 2 FIRMS

PROFIT AND LOSS STATEMENT (×1,000)	COMPANY 1	COMPANY 2
Net Sales Billed	48,000.	48,000.
Direct Cost of Sales	30,000.	31,000.
Depreciation	250.	250.
Gross Margin	17,750.	16,750.
Commercial and Administrative:		
Advertising	4,000.	9,000.
Inventory Charges	2,750.	2,500.
Administrative Overheads (Fixed)	500.	500.
Interest Expense	0	0
Net Profit (−Loss)	10,500.	4,750.
Tax Reserve	5,250.	2,375.
Net Profit (−Loss) After Taxes	5,250.	2,375.
Dividends Paid	0	0
Addition to Net Worth	5,250.	2,375.

BALANCE SHEET (×1,000)	COMPANY 1	COMPANY 2
Short Term Assets		
Cash	14,750.	19,375.
Inventories (at Cost)	45,000.	37,500.
Total	59,750.	56,875.
Long Term Assets		
Plant	10,000.	10,000.
Total Assets	69,750.	66,875.
Liabilities		
Loans Outstanding	0	0
Net Worth	69,750.	66,875.
Line of Credit	20,000.	20,000.

Direct cost of sales includes the actual variable costs of changing production. One has the choice of charging inventory costs directly to sales or including them under administrative costs. This is a matter of convention.

Depreciation is the amount to be charged against the deterioration of the plant. The size and meaning of depreciation in this model has been selected to conform with relatively simple economic theorizing. There is however considerable leeway in practice. The method used here can be described briefly as follows: imagine that the amount paid in depreciation is precisely the amount required to maintain the plant at its top capacity in its original condition. At any point in the game the plant will be exactly the same as it was at any other point in the game since depreciation is charged every quarter. The exception of course is if the players are permitted to add new capacity, in which case depreciation charges per quarter will rise, but the same policy is followed. This means that we have removed from consideration the possibility of a whole set of important policies devoted to running down or building up plant.

The gross margin is obtained by subtracting the direct cost of sales and the depreciation from net sales billed. Commercial and administrative costs are costs that are not allocated directly to each item sold. As has been noted above accounting conventions give the individual an option in some instances. The option is available since accounting is used for many purposes and depending upon the specific purpose and industry the categorization of costs may vary. An accounting system is a model for use in a decision system. As the uses vary, so the optimum design of the system varies.

Advertising is represented by the dollar cost of the amount spent during each quarter.[3] The effect of advertising is illustrated in Chapter 11.

Inventory charges are the cost of carrying a unit of inventory, an important parameter in this model. Inventory charges are levied on the number of units of inventory carried on the average over the period. We take the initial inventory and the ending inventory divided by two and multiply this by the unit inventory carrying cost.

Administrative overheads cover the payments to personnel not directly engaged in production. These include the officers of the company, administrative staff, and others. Once more it can be seen that the definition of these expenses permits a great amount of leeway.

If the firm has bank debts or other instruments of debt such as bonds it

3. There is a special problem with the classification of advertising and sales effort. In some industries it is reasonable to regard them as current expenditures, but in many instances they can more accurately be regarded as investments in building the firm's marketing organization. This means that when a firm is reporting losses, in fact it may be in very good shape and investing heavily.

must pay interest. These payments will appear as an expense. In this market model we do not consider bonds, but do take into account the possibility that the firm may borrow from the bank. In the above profit and loss statement there are no interest charges since there is no debt during the second quarter.

Subtracting the commercial and administrative costs from the gross margin gives the net profit before taxes. Taxes must be subtracted before net profit is obtained. We assume that taxes are at the rate of 50%. This is not necessarily so and a parameter can be adjusted to reflect the going rate.

After dividends have been paid, the remaining profit constitutes an addition to the net worth of the firm and is so recorded.

In the financial report an extremely simple balance sheet is given. The first item under short-term assets is cash on hand. The second item is inventories evaluated at direct variable cost. This understates the cost of production of the inventories. If the firm is healthy this will considerably understate the value of inventories as an asset to the firm. If the firm has made errors, the inventories could be a white elephant with a market value of far less than their cost of production. If we wished to be ultraconservative we could value inventories as the minimum of cost or market value.

In many markets costs of production change. This may cause a firm to have an inventory of like items that cost different amounts to produce. When these items are sold, reported profits are affected if the firm claims to be supplying the market with high or low-cost inventory items. Tax considerations require that a firm select and stick to a specific convention, either LIFO or FIFO. In the first instance the last item in inventory is presumed to be the first item out; thus in a market with rising costs it will underestimate a firm's profits. In the second system, the first item in inventory is presumed to be the first out; hence with rising costs, profits are overestimated. Different results from different conventions stress the need for sophistication in reading even simple company reports and call for an understanding of the difficulties in obtaining good models of dynamic economic processes.

Plant is the only long-term asset in this simple model. Depreciation charges are assumed sufficient to maintain plant at a constant worth (or at a larger worth if investment is made to enlarge plant). If the cost of new plant is changing, as is often the case, a host of difficult problems must be faced by management in evaluating their assets and establishing adequate funds for replacement and expansion. With inflation of prices in items such as land, corporate balance sheets will frequently be grossly conservative in their valuation of plant. For example not long ago several film companies carried their holdings of the land values of their studios located in the heart of Los Angeles at their purchase price of many decades ago.

The liability side of a balance sheet usually carries entries for short and long-term liabilities. In financial welfare a firm may become vulnerable by being caught in a cash squeeze from large short-term liabilities. It is for this reason that cash flow is of such great concern to many firms. In our simplified model, the financial features are rudimentary. Each firm is regarded as having a revolving line of credit at the bank. In other words, whenever the firm needs credit (up to a certain limit) it obtains it automatically and can repay at any time. The firm only runs the risk of bankruptcy when it has exhausted its line of credit. In a world with imperfect money markets and difficulties in evaluating risk and the credit worthiness of firms our treatment here is ideal in the extreme.

The only liabilities in this model are the loans outstanding. The loans outstanding plus the equity of the owners amount to the total assets of the firm.

The following is given to each firm to enable the management to tally its score:

QUARTER 2, GAME 1A, 2 FIRMS

	Company 1	Company 2
Value of Accumulated Dividends ($000)	0	0
Current Net Worth	69,750.	66,875.
Total	69,750.	66,875.

This display gives the value of the accumulated dividends, which includes not only the dividends but the interest earned upon them, assuming that they have been invested at the going rate; and the net worth of the firm. These entries provide a total by which management may judge its overall economic performance.

Accounting and Decision Systems

Only the most rudimentary of accounts have been presented here so that they can be seen in relation to the economic structure given in Chapters 5 and 6 and elaborated in subsequent chapters. Accounting involves an aggregation and presentation of financial information about the firm. Even at this level of simplicity, accounting serves several goals (and those goals may conflict with each other).

Accounting is used for tax purposes. It is used to obtain an overall evaluation of the worth of a firm. This is operationally important to groups who are contemplating buying or selling firms. Accounting may be used for internal control and as a basis for production, marketing, investment, and financial decisions. It is highly unlikely that a single set of accounting numbers can serve all of these purposes simultaneously with high efficiency. The possible incompatibility of the different needs has caused many legal, behavioral, and economic problems in the running of corporations. This is especially true in the case of large multiproduct, multiplant, international firms who attempt to use their accounting systems to decentralize and rationalize decisionmaking.

The models in many parts of this book are extreme abstractions, useful to us because they permit economic theorizing and analysis. They lack institutional detail and avoid arbitrary social conventions when unnecessary. The models themselves do not provide the viewing devices needed to link theory with actual markets and firms. The understanding of an accounting system is a means to link the theory to the institution. Taxes, legal and accounting conventions, crude approximations of complex technology are, in general, neither interesting nor elegant. However, an economist wishing to apply his knowledge to an economic institution must be able to comprehend and appreciate these details, if he is to avoid being led utterly astray by overlooking dull but vital details.

Economics and Generally Accepted Accounting Principles

Not only is accounting part art and a highly applied endeavor, it is also based on an important dynamic mixture of law, custom, and consensus. The microeconomist wishing to understand how economic facts are portrayed in the accounting figures is well advised to investigate the basic premises on which the accounting system he uses was built. In the United States, the microeconomist who uses the financial reporting of corporations should appreciate Generally Accepted Accounting Principles (Am. Inst. of Certified Public Accountants, 1971, Section 1025, 1026).

Thirteen statements and three modifying conventions suffice to illustrate the degree of approximation, the elements of compromise, and the bias toward conservatism in current financial accounting.

The thirteen statements are:

1. *Accounting entity.* Accounting information pertains to entries, which are circumscribed areas of interest. In financial accounting the entity is the specific business enterprise. The enterprise is identified in its financial statements.
2. *Going concern.* An accounting entity is viewed as continuing in operation in the absence of evidence to the contrary.[4]
3. *Measurement of economic resources and obligations.* Financial accounting is primarily concerned with measurement of economic resources and obligations and changes in them.

4. The corollary observation is that if liquidation seems imminent, financial information may be prepared on the assumption that liquidation will occur.

4. *Time periods*. The financial accounting process provides information about the economic activities of an enterprise for specified time periods that are shorter than the life of the enterprise. Normally the time periods are of equal length to facilitate comparisons. The time period is identified in the financial statements.
5. *Measurement in terms of money*. Financial accounting measures monetary attributes of economic resources and obligations and changes in them. The unit of measure is identified in the financial statements.
6. *Accrual*. Determination of periodic income and financial position depends on measurement of economic resources and obligations and changes in them as the changes occur rather than simply on recording receipts and payments of money.
7. *Exchange price*. Financial accounting measurements are primarily based on prices at which economic resources and obligations are exchanged.
8. *Approximation*. Financial accounting measurements that involve allocations among relatively short periods of time and among complex and joint activities are necessarily made on the basis of estimates.
9. *Judgment*. Financial accounting necessarily involves informed judgment.
10. *General-purpose financial information*. Financial accounting presents general-purpose financial information that is designed to serve the common needs of owners, creditors, managers, and other users, with primary emphasis on the needs of present and potential owners and creditors.
11. *Fundamentally related financial statements*. The results of the accounting process are expressed in statements of financial position and changes in financial position, which are based on the same underlying data and are fundamentally related.
12. *Substance over form*. Financial accounting emphasizes the economic substance of events even though the legal form may differ from the economic substance and suggest different treatment.
13. *Materiality*. Financial reporting is only concerned with information that is significant enough to affect evaluations or decisions. The three modifying conventions are:

Conservatism

Frequently, assets and liabilities are measured in a context of significant uncertainties. Historically, managers, investors, and accountants have generally preferred that possible errors in measurement be in the direction of understatement rather than overstatement of net income and net assets. This has led to the convention of conservatism, which is expressed in rules adopted by the profession as a whole such as the rules that inventory should be measured at the lower of cost and market that accrued net losses should be recognized on firm purchase commitments for goods for inventory. These rules may result in stating net income and net assets at amounts lower than would otherwise result from applying the pervasive measurement principles.

Emphasis on Income

Over the past century businessmen, financial statement users, and accountants have increasingly tended to emphasize the importance of net income and that trend has affected the emphasis in financing accounting. Although balance sheets formerly were presented without income statements, the income statement has in recent years come to be regarded as the most important aspect of the financial statements. Accounting principles that are deemed to increase the usefulness of the income statement are therefore sometimes adopted by the profession as a whole regardless of their effect on the balance sheet or other financial statements.

Application of Judgment by the Accounting Profession

Sometimes strict adherence to the pervasive measurement principles produces results that are considered by the accounting profession as a whole to be unreasonable in the circumstances or possibly misleading. Accountants approach their task with a background of knowledge and experience. The perspective provided by this background is used as the basis for modifying accounting treatments when strict application of the pervasive measurement principles yields results that do not appear reasonable to the profession as a whole.

Appendix B | Contingent Demand and Rationing

THIS APPENDIX provides both the general theoretical basis for the description of contingent demand and is needed for the algorithm used to assign demand to the players in an oligopoly game based on the theory in this volume. The problem of the assignment of contingent demand is closely related to rationing schemes in general disequilibrium analysis. In a closed economic system in several sectors how does rationing take place and who absorbs the inventory losses if there is excess supply or demand? The analysis provided here[1] answers these questions.

The discussion of contingent demand in the text may be regarded as a special case of what is presented here. The level of mathematics in this Appendix is considerably higher than in the text of this book. The reader more interested in oligopoly models may wish to skim or bypass this Appendix. The notation here is independent of the text.

For more than a century, economists have been interested in the behavior of oligopolists. A theory of oligopoly requires assumptions about the demand function for the product(s) of each firm. More recently, students of market research have been interested in the demand function facing an oligopolistic firm. For the economist and the market-research analyst, a demand function is an empirical relationship concerning the purchasing behavior of a group of buyers. However, before such an empirical relationship can be ascertained, by econometric investigations

1. Appendix B is based heavily upon the work of Levitan (1964).

or by experiments with business games, a model of demand must be formulated. This Appendix deals with the formulation of mathematical models of demand facing oligopolists. It deals with general properties of oligopoly demand functions, as implicit in some rather weak assumptions about the structure of demand.

In this Appendix demand for the product of an oligopolistic firm is assumed to be a function of the prices of all firms in the industry; the main question is the construction of non-negative demand functions. Following this, the quantities offered by all firms, in addition to the prices set by them, are allowed to influence the quantities purchased, since unfilled demand for one product may lead to demand for a substitute. Also, the question is posed and partially answered of whether a function describing demand as a function of prices alone for those values of the prices that give non-negative values of the function can be extended in a natural way to describe demand when there are supply limitations and for all non-negative prices.

An illustrative numerical example for a two-good case is presented which follows with sufficient conditions for a function to have a unique extension. The structure of the proof of a unique extension is similar to proofs of the existence of equilibrium in competitive economic systems.

Implications of a subset of the unique extension postulate on the value of the parameters of a linear demand function are then examined. These postulates permit an interesting interpretation of the parameters.

In the concluding sections, we examine the implications for a linear market of the products being divided into subgroups of close substitutes. It is shown that if the grouping obeys the postulates of Pearce's schema, there is implied a distinct simplification of the parameter structure. Finally, we discuss computational schemes for extending numerical demand functions in the way required above. It is shown that in the linear case the problem can be solved by a special finite algorithm. For the more general case, a convergent algorithm is given.

Price Demand Functions

A demand function for products of oligopolists is a function that maps the set of possible market strategy choices of the oligopolists into the set of possible purchasing decisions of the buyers in the market. We consider a market in which n products are sold, and we let the strategy space be the ordered set of pairs of price-supply decisions. Our aim is to derive a simple coherent model of the relationship of sales to the set of price and quantity decisions of sellers.

In the standard model, demand is not considered a function of supply quantities, but a function of price alone; this will adequately represent demand if the quantities demanded do not exceed supplies offered. Here we treat demand as function of prices alone. Specifically, we are interested in the question of whether the functional form representing a demand relationship can give zero and negative values for quantities demanded at economically meaningful values for prices. In the following section, the effect of limited supply quantities will be taken up.

Negative demand means that at certain values of price, the economic agents in the buying market would offer on balance some quantity of a product for sale. If there are institutional or technological reasons that maintain the sellers as monopolists of their own product, demand is certainly restricted to non-negative. If there is no such constraint, the negative sales by sellers will not take place unless the seller's offer is to sell or buy at his offer price. We will assume that our sellers are not making buying offers at their published prices. For these reasons, it is clear that demand will be non-negative.

Now, it may be difficult to find simple functional forms that are non-negative over the non-negative price orthant, R_n^+, and that will represent demand functions. Constant elasticity functions $d(p) = \{k_i \Pi p_j^{\eta_{ij}}\}$ have been used in econometric investigations. However, it is doubtful that in many markets a seller could expect to have positive demand at every price combination and have unlimited sales at very low prices.

It is desirable, therefore, to be able to ignore the non-negativity condition in choosing a functional form and to be able to derive from it another function that equals the original function in that part of the price space where the former is non-negative, but which is modified in that part in which the chosen function has negative components. In order to make such a derivation, we will postulate two properties for demand functions. We shall incorporate these postulates into a formal definition.

Definition 1. An n-good *price demand function* is a function, f, that maps R_n^+ (the space of ordered n-tuples of non-negative real numbers) into R_n^+, such that if $f_i(\bar{p}) = 0$ and $\Delta \geq 0$, then $f(\bar{p} + \Delta\delta_i) = f(\bar{p})$ where δ_i is the ith unit vector.

The meaning of this postulate is fairly simple: if, at some set of prices, there is no demand for a particular good, an increase in the price of that good will have no effect on the quantities demanded of all goods. This is a fairly reasonable assumption based on the nonexistence of Veblen effects of posted prices on goods that are unsold on the consumption of substitute goods. We will use this postulate to derive a demand function from a more general function that does not embody the non-negativity constraint. We do this by defining a problem.

Problem 1. Given a function $f: R_n^+ \rightarrow R_n$ and a vector $p \in R_n^+$, find a vector $p' \in R_n^+$, such that $f(p') \in R_n^+$ and $p'_i = p_i$ when $f_i(p') > 0$ and $p'_i \leq p_i$ when $f_i(p') = 0$.

Theorem 1. *Let $f = R_n^+ \to R_n$. If for every $p \in R_n^+$, Problem 1 has a unique solution $p'(p)$, then $\hat{f}: R_n^+ \to R_n^+$, where $\hat{f}(p) = f(p'(p))$ is a unique n-good price demand function that agrees with f for those values of p for which $f(p) \in R_n^+$.*

Proof. Let $\bar{p} \in R_n^+$, such that $\hat{f}_i(\bar{p}) = 0$; and let $\bar{\bar{p}} = \bar{p} + \Delta\delta_i$ where $\Delta > 0$. By assumption, there exists a unique p', such that $f(p') \in R_n^+$, $f_j(p') > 0$ implies $p'_j = \bar{p}_j$, and $f_j(p') = 0$ implies $p'_j \leq \bar{p}_j$. By the choice of index i, it is apparent that p' solves Problem 1 for f and $\bar{\bar{p}}$. By the uniqueness assumption $p'(\bar{\bar{p}}) = p'(\bar{p})$, $f(\bar{\bar{p}}) = f(\bar{p})$, and hence $\hat{f}$ is a price demand function.

Suppose ϕ is a price demand function that agrees with f on $S = \{p \mid f(p) \in R_n^+\}$. Let $p \notin S$. By assumption there exists a unique $p' \leq p$ such that $f(p) \geq 0$ and $p'_i < p_i$ implies $f_i(p) = 0$. Let $J = \{i \mid p'_i < p_i\} = \{j_1, \ j_2, \ldots, j_k\}$. Let $p^{(0)} = p'$, $p^{(l)} = p^{(l-1)} + (p_{j_l} - p'_{j_l})\delta_{j_l}$. Since ϕ is a price demand function, $\phi(p^{(l)}) = f(p') = \hat{f}(p)$ for $l = 1, 2, \ldots, k$. But $p^{(k)} = p$ and the theorem is proved.

The effect of the construction of the function $\hat{f}$ in Theorem 1 is that we solve for a set of new prices for goods that have no sales which just serves to make demand zero for these goods. It will be useful to call the transformed price vector p', the effective price vector.[2] For goods that have positive sales its components are the transaction prices; for each good with no sales its component is (if f_i is strictly decreasing in p_i) precisely the least upper bound on the set of prices at which the good can be sold. We defer the discussion of conditions on functions that ensure the unique existence of solutions to Problem 1.

Price-Quantity Demand Functions

We now turn to the problem of representing the effect of supply shortages on demand. As mentioned above, the classical price demand function gives demand under the assumption that quantities of goods are freely available in sufficient quantity to satisfy the requirements of all consumers at the prices established. In the market we wish to study, at least in the short run, some seller may not make enough goods available to satisfy the purchasers. Presumably, the effect of such a shortage will be to divert purchases to substitute goods. What is required in this case is some analysis that would extend the concept of a price demand function to a price-quantity demand function which would give sales as a function of not only prices but also of supply availabilities.

Shubik (1959b) and Shubik and Levitan (1962) have proposed models of markets in which short goods are rationed by various mechanisms and demand functions are derived for customers whose wants are not com-

2. For a version of this concept, see Papandreou and Wheeler (1954, pp. 38–41).

pletely satisfied. We shall propose here a very simple mechanism for this function, that is, the market acts as though it contained some very efficient arbitrageur who can establish a price that rations the available supply of short goods. Such an assumption will be seen to contain no additional quantitative assumptions beyond those embodied to our device for extending general functions to price demand functions.

Definition 2. An n-good *price-quantity demand* function is a function $f: R^+_{2n} \to R^+_n$ such that

(1) $f(p, q) \leq q$.

(2) For each $p \in R^+_n$, there exists a unique $\bar{q}(p)$ such that if $q \geq \bar{q}(p)$, $f(p, q) = \bar{q}(p)$ and $\bar{q}(p)$ is an n-good price demand function.

(3) If $f_i(p, q) = q_i$ and $\Delta \geq 0$, then $f(p - \Delta\delta, q) = f(p, q)$.

The first two postulates of the definitions are clearly required for a demand function. The last, that the lowering of the price of a short good has no effect on demand, is questionable. It assumes that either there are no income effects in consumption or that the arbitrageurs dispose of their income in such a way as to keep it insulated from the market. We shall only justify this postulate by pointing out that it will enable us to avoid complicated institutional assumptions about rationing and demand diversion. The development of a quantity-price demand function from a price-demand function follows from the following problem and theorem.

Problem 2. Given an n-good price demand function f and vector $p, q' \in R^+_n$, find a vector p' such that $f(p') \leq q$, $p'_i = p_i$ when $f_i(p') < q_i$; $p'_i \geq p_i$ when $f_i(p) = q_i$.

Theorem 2. If for a given function, f, Problem 2 has a solution $p'(p, q)$ for every $(p, q) \in R^+_{2n}$, which is unique except for those components corresponding to $f_i(p') = 0$, then $f(p'(p, q))$ is a unique price-quantity demand function that agrees with $f(p)$ for those values of p such that $f(p) \leq q$.

The proof of Theorem 2 is exactly analogous to that of Theorem 1.

We have shown that if Problem 2 has a solution there exists for all price-quantity combinations a set of derived or effective prices p' that serve to ration the scarce goods and allocate demand to those with excess supply. From a functional form that describes demand when all demands are non-negative, we can derive a unique price-quantity demand function that is consistent with our original function assuming that Problems 1 and 2 have unique solutions.

For further purposes it will be useful to combine Problems 1 and 2.

Problem 3. Given a bounded function $f: R_n^+ \to R_n$ and vector $p, q \in R_n^+$. Find a vector p' such that $0 \leq f(p') \leq q$; if when $f_i(p') = 0$, $p_i' \leq p_i$, when $0 < f(p') < q_i$, $p_i' = p_i$ and when $f_i(p') = q_i$, $p_i' \geq p_i$.

Theorem 3. If for a given function f and for every $(p, q) \in R_{2n}^+$ there is a unique solution $p'(p, q)$ to Problem 3, then

(1) *Problem 1 has a unique solution $\bar{p}(p)$ for f and every p.*

(2) *Problem 2 has a solution $\bar{\bar{p}}(p, q)$ for $f(\bar{p}(p))$ and every p and q which is unique except for those components for which $f_i(\bar{p}) = 0$.*

(3) *$p'(p, q)$ solves the problem of Part 2 and $f(p'(p, q))$ is a unique price-quantity demand function which extends f.*

Proof. Part (1) follows immediately if we let $q = (\bar{q}_i)$, a set of upper bounds on the components of f. Let $\bar{f}(p) = f(\bar{p}(p))$, and p^0, $q^0 \in R_n^+$. By hypothesis, Problem 3 has a unique solution p' for p^0, q^0, and f. By Part (1), there is a unique $\bar{p}(p')$ such that $f(\bar{p}(p')) \geq 0$; $f_i(\bar{p}(p')) > 0$ implies $p_i(p') = p_i'$ and $f_i(p') = 0$ implies $\bar{p}_i(p') \leq p_i'$. But p_i' satisfies these conditions hence $\bar{p}(p') = p'$ and $f(\bar{p}(p')) = f(p')$. By assumption $f(\bar{p}(p')) = f(p') \leq q^0$ and $f_i(p') < q_i$ implies $p_i' = p_i^0$ and $f_i(p') = q_i$ implies $p_i' \geq p_i^0$ hence p' is a solution of the problem of Part (2). Suppose p'' is a solution to the problem of Part (2). Then $f(\bar{p}(p'')) \leq q$; $f_i(\bar{p}(p'')) < q_i$ implies $p_i'' = p_i^0$; $f_i(\bar{p}(p')) = q_i$ implies $p_i'' \geq p_i^0$. By the construction of $\bar{p}(p)$, $f(\bar{p}(p'')) > 0$ implies $\bar{p}_i(p'') \leq p_i''$. Hence we have $\bar{p}(p'')$, which is a solution of Problem 3, hence $\bar{p}(p'') = p'$ and $f_i(\bar{p}(p'')) > 0$ implies $p_i'' = p_i'$.

A Numerical Duopoly Example

To illustrate the ideas of the previous section we give a numerical example of a symmetric, differentiated duopoly demand function. Let $q = R_2^+ \to R_2$ be given by

$$q(p) = (q_1(p), q_2(p)) =$$

$$\begin{cases} \begin{pmatrix} 800 - 320p_1 + 200p_2 + (p_1 + p_2)^2 \\ 800 + 200p_1 - 320p_2 + (p_1 + p_2)^2 \end{pmatrix} & \text{for } \begin{pmatrix} 0 \\ 0 \end{pmatrix} \leq p \leq \begin{pmatrix} 10 \\ 10 \end{pmatrix} \\ \begin{pmatrix} -1 \\ -1 \end{pmatrix} & \text{otherwise} \end{cases}$$

This example, with two products, is convenient because we shall show that the function q will give rise to unique solutions of Problem 3 without the combinatorial difficulties that beset problems with more products. We will be able to show graphically the contours of the price-demand function and the demand for one product as a function of prices and quantities.

We start with the key existence and uniqueness proposition.

Proposition 1. Given the function q as defined above and vectors p and s in R_2^+, there exists a unique solution p' in R_2^+ to Problem 3 (that is, $0 \leq q(q') < s$, $p_i' > p_i$ implies $q(p') = s_i$, and $p_i' \in p_i$ implies $q_i(p') = 0$).

Proof. The function $\phi(x, y) = 160 - x - \sqrt{24800 - 520x + y}$ is the solution, for those values of (x, y) where $\phi(x, y) \geq 0$ to the problems $q_1(\phi, x) = y$ or $q_2(x, \phi) = y$. We let $\theta(x, y) = 15 + (y - x)/1040 - \sqrt{25 + (x + y)/8}$ and

$$u = (\theta(s_1, s_2), \theta(s_2, s_1))$$
$$v = (\theta(0, s_2), \theta(s_2, 0))$$
$$w = (\theta(s_1, 0), \theta(0, s_1)).$$

The points u, v, and w respectively are in R_2^+ and are solutions to $q(u) = (s_1, s_2)$, $q(v) = (0, s_2)$, and $q(w) = (s_1, 0)$ if and only if the respective problems have solutions in R_2^+. Further we note that $q(10, 10) = (0, 0)$ *or* $\theta(0, 0) = 10$.

We partition R_2^+ in the following mutually disjoint sets, some of which, depending on s may be void.

$$S_1 = \{x | x = (x_1, x_2); \max[0, \phi(x_2, s_1)] \leq x_1 \leq \phi(x_2, 0); \max[0, \phi(x_1, s_2)] \leq x_2 \leq \phi(x_1, 0)\}$$
$$S_2 = \{x | 10 < x_1, 10 < x_2\}$$
$$S_3 = \{x | 0 \leq x_1 < u_1, 0 \leq x_2 < u_2\}$$
$$S_4 = \{x | \max(0, v_1) \leq x_1; 0 \leq x_2 \leq v_2\}$$
$$S_5 = \{x | 0 \leq x_1 \leq w_1; \quad \max(0, w_2) \leq x_2\}$$
$$S_6 = \{x | \phi(x_2, 0) < \phi_1; \max(0, v_2) \leq x_2 \leq 10\}$$
$$S_7 = \{x | \max(0, w_1) \leq x_1 \leq 10; \phi(x_1, 0) < x_2\}$$
$$S_8 = \{x | \max(0, u_1) \leq x_1 \leq v_1; 0 \leq x_2 < \phi(x_1, s_2)\}$$
$$S_9 = \{x | 0 \leq x_1 < \phi(x_2, s_1); \max(0, u_2) \leq x_2 < w_2\}.$$

These sets are illustrated in Figure A.1, with $s = (400, 800)$.

We shall now demonstrate a solution p' to problem 3 for p in each of the sets S_i, where $i = 1, \ldots, 9$.

$$p'(p, s) = \begin{cases} p & \text{if } p \in S_1 \\ (10, 10) & \text{if } p \in S_2 \\ u & \text{if } p \in S_3 \\ v & \text{if } p \in S_4 \\ w & \text{if } p \in S_5. \\ (\phi(p_2, 0), p_2) & \text{if } p \in S_6 \\ (p_1, \phi(p_1, 0)) & \text{if } p \in S_7 \\ (p_1, \phi(p_1, s_2)) & \text{if } p \in S_8 \\ (\phi(p_2, s_1), p_2) & \text{if } p \in S_9 \end{cases}$$

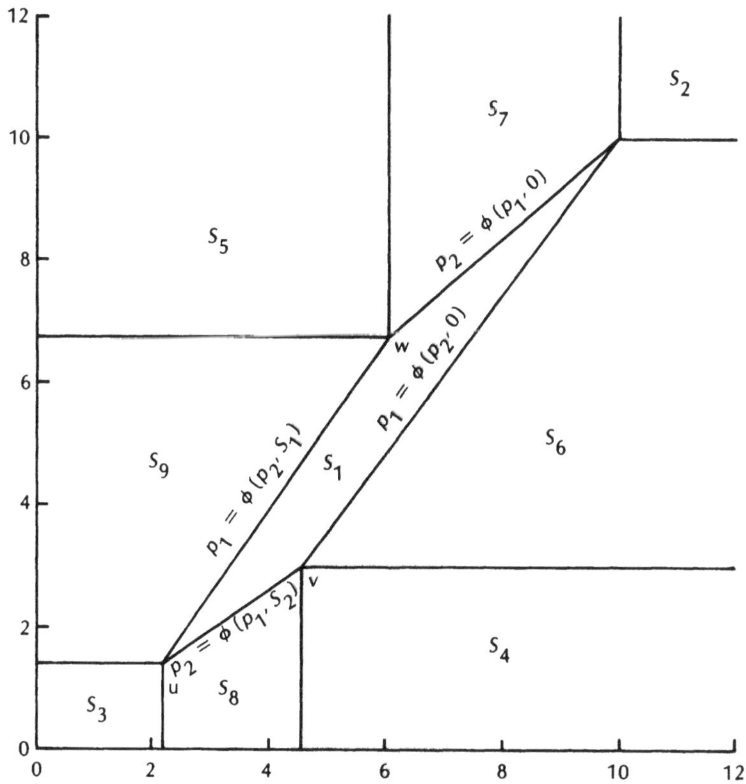

Figure A.1. Regions in the calculation of contingent demand.

It is apparent by detailed check of the function $p'(p, s)$ that it is indeed a solution to Problem 3.

For a given p and s suppose there are two solutions, $\bar{p} \neq \bar{\bar{p}}$. Without loss of generality we may assume that $\bar{\bar{p}}_1 - \bar{p}_1 \geq |\bar{\bar{p}}_2 - \bar{p}_2|$ and $\bar{\bar{p}}_1 > \bar{p}_1$. Now either $\bar{p}_1 < p_1$ or $\bar{\bar{p}}_1 > p_1$ so either $q_1(\bar{\bar{p}}) = s_1$ and $0 \leq q_1(\bar{p}) \leq s_1$ or $q_1(\bar{p}) = 0$ and $0 \leq q_1(\bar{\bar{p}}) \leq s_1$ so that $q_1(\bar{\bar{p}}) - q_1(\bar{p}) \geq 0$.

Now,

$$\begin{aligned}
q_1(\bar{\bar{p}}) - q_1(\bar{p}) &= -320(\bar{\bar{p}}_1 - \bar{p}_1) + 200(\bar{\bar{p}}_2 - \bar{p}_2) + (\bar{\bar{p}}_1 + \bar{\bar{p}}_2)^2 - (\bar{p}_1 + \bar{p}_2)^2 \\
&= (-320 + \bar{\bar{p}}_1 + \bar{p}_1 + 2\bar{\bar{p}}_2)(\bar{\bar{p}}_1 - \bar{\bar{p}}_1) + (200 + \bar{\bar{p}}_2 \\
&\quad + \bar{p}_2 + 2\bar{p}_1)(\bar{\bar{p}}_2 - \bar{p}_2) \\
&\leq (-320 + \bar{\bar{p}}_1 + \bar{p}_1 + 2\bar{\bar{p}}_2)(\bar{\bar{p}}_1 - \bar{p}_1) + (200 + \bar{\bar{p}}_2 \\
&\quad + \bar{p}_2 + 2\bar{p}_1)(\bar{\bar{p}}_1 - \bar{p}_2) \\
&= (-120 + \bar{\bar{p}}_1 + 3\bar{p}_1 + 3\bar{\bar{p}}_2 + \bar{p}_2)(\bar{\bar{p}}_1 - \bar{p}_1) \\
&\leq -40(\bar{\bar{p}}_1 - \bar{p}_1) \\
&< 0,
\end{aligned}$$

since $\bar{\bar{p}}_1, \bar{\bar{p}}_2, \bar{p}_1, \bar{p}_2 < 10$. This is a contradiction; hence uniqueness is proved.

Corollary. There exists a unique price demand function $\hat{q}(p) = q(p'(p(10, 10))$ *and a unique price-quantity demand function,* $\hat{\hat{q}}(p, q)$ *which agrees with* q.

In Figure A.2 we show some contours of $\hat{q}_1$ and $\hat{q}_2$. The sloping curves are functions of the form, $p_i = \phi(p_j, \hat{q}_i)$.

There is no simple graphical way to represent the function $\hat{\hat{q}}$. Figure A.3 represents a set of graphs of $\hat{\hat{q}}_1(p_1, p_2, 1400, q_2)$ when p_2 has values 0, 400, 800, and 1200. For example, when $p_2 = 4$ and $q_2 = 400$, the graph of $\hat{q}_1(p_1)$ is a three-branched curve consisting on the left of the dotted line

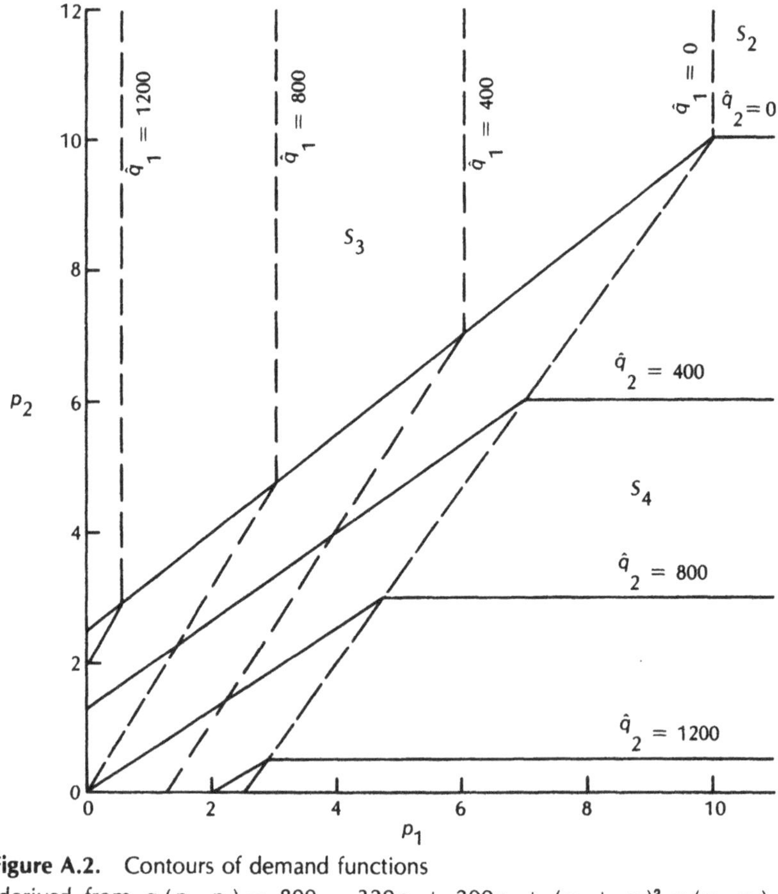

Figure A.2. Contours of demand functions (derived from $q_1(p_1, p_2) = 800 - 320p_1 + 200p_2 + (p_1 + p_2)^2$ $q_2(p_1, p_2) = 800 + 200p_1 - 320p_2 + (p_1 + p_2)^2$).

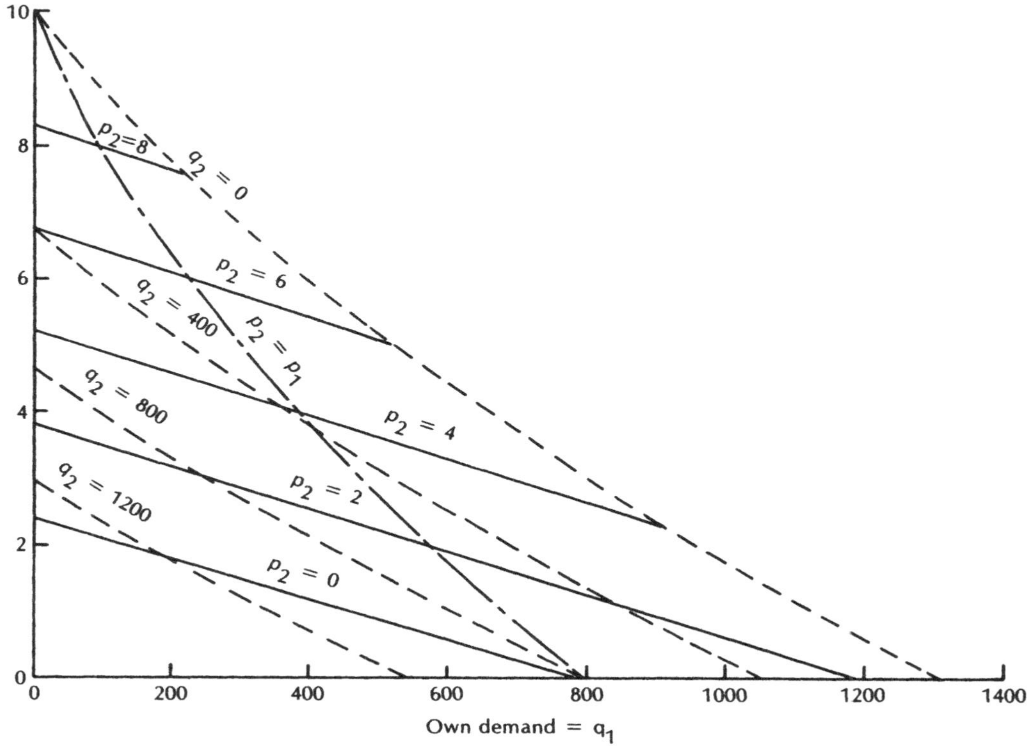

Figure A.3. Duopoly demand as a function of own price (when other price is constant: ———; other sales are constant: ––––; other price equals own price: — - —).

labeled $q_2 = 400$, up to its intersection with the solid line labeled $p_2 = 4$; the graph follows this curve up to the intersection with $q_2 = 0$ and follows it until $\hat{p} = 0$. These three branches represent values of p_1 for which product 2 is in short supply, is in available supply and selling positive quantities, and finally is priced out of the market. We call this curve the demand for product one conditioned on the price and supply of product two. As may be seen by inspection, depending on (p_2, q_2) the conditional demand may be a one, two, or three branched curve.

General Sufficient Conditions for Existence and Uniqueness

The aim of this section is to present some plausible conditions for there to exist solutions to Problem 3 and conditions for these solutions to be unique.

We will begin with the existence conditions.

Postulate 1. *f is a continuous function.*

Postulate 2. $S = \{p | p \in R_n^+, f(p) \in R_n^+\}$ *is bounded.*

Postulate 3. *Let* $\bar{p}_i = \max_{p \in S} p_i$ *and let* $T = \Pi[0, \bar{p}_i]$. *(Π means Cartesian product)*

(a) $p \in T$ *and* $p_i = 0$ *implies* $f_i(p) \geq 0$

(b) $p \in T$ *and* $p_i = \bar{p}_i$ *implies* $f_i(p) \leq 0$.

Postulate 1 is noncontroversial. Postulate 2 says there is an upper bound to the price at which each good can be sold. Postulate 3 says that there is a non-negative demand for each good at zero price, and a nonpositive demand at the maximum price if all other prices are in the respective ranges at which demand is non-negative.

Theorem 4. *If the function* $f: R_n^+ \to R_n$ *satisfies Postulates 1, 2, and 3, then, given any p and s in* R_n^+, *Problem 3 has a solution,* p^0; *that is, there exists a* p^0 *such that* $0 \leq f(p^0) \leq s$ *and* $p_i^0 > p_i$ *implies* $f_i(p^0) = s_i$ *and* $p_i^0 < p_i$ *implies* $f_i(p^0) = 0$.

Proof. Let $\hat{p} = (\min(p_i, \bar{p}_i))$ and consider the mapping $H: R_n^+ \to 2^T$ given by $H(x) = \Pi_i H_i(x)$ where

$$H_i(x) = \begin{cases} \{0\} & \text{if } f_i(x) < 0 \\ \{0, \hat{p}_i\} & \text{if } f_i(x) = 0 \\ \{p_i\} & \text{if } 0 < f_i(x) < s_i. \\ [\hat{p}_i, \bar{p}_i] & \text{if } f_i(x) = s_i \\ \{\bar{p}_i\} & \text{if } f_i(x) > s_i \end{cases}$$

It is apparent that $H(x)$ is compact, and maps T into 2^T.

Consider a sequence (x^k) such that $\lim x^k = x^0$ and a sequence (y^k) such that $\lim y^k = y^0$ and $y^k \in H(x^k)$. We assert that $y^0 \in H(x^0)$ and hence H is upper semicontinuous.

For any i either
(a) $f_i(x^0) < 0$, $0 < f_i(x^i) < s_i$, or $f_i(x^0) > s_i$; or
(b) $f_i(y^0) = 0$ or $f_i(x^0) = s_i$.
In case (a), by the continuity of f all but a finite set of the sequence $(f_i(x^k))$ is in the respective open set and the corresponding values y^k and equal respectively to 0, $\hat{p}_i$, or $\bar{p}_i$, and hence so does y^0. In case (b), $H_i(x^0)$ is either $[0, \hat{p}_i]$ or $[\hat{p}_i, \bar{p}_i]$ and by closure must contain any limit of points in it. Thus in both cases $y^0 \in H(x^0)$ and the assertion is proved.

Thus we may invoke the Kakutani fixed-point theorem, and assert that there exists a point p^0 such that $p^0 \in H(p^0) \in T$. For any i, $f_i(p^0) < 0$ or $f_i(p^0) > s_i$ by the construction of H contradicts Postulate 3. Further $p_i^0 < p_i$ implies $p_i^0 < \hat{p}_i$ which implies $f_i(p^0) = 0$, and $p_i^0 > p_i$ implies $p_i^0 > \hat{p}_i$ which implies $f_i(p^0) = s_i$.[3]

We now turn to the question of uniqueness of the solution. We make one additional postulate here.

Postulate 4. f is differentiable over S^c, the convex hull of S.
(a) $(\partial/\partial p_j)f_i(p) > 0$ *if* $i \neq j$ *and* $p \in S^c$
(b) *There is a vector* $q \in R_n^+$, *such that* $q_i > 0$, *and* $\Sigma_j q_i(\partial/\partial p_j)f_i(p) < 0$ *for* $i = 1, 2, \ldots, n$, *and* $p \in S^c$.

It should be noted that Postulate 4(a), if taken over T, together with the condition, $0 \in T$, implies Postulate 3. This part says that all goods are gross substitutes in consumption. The second part says the Jacobian matrix of partial derivatives is of the Leontief type. In economic terms it means that there is a set of non-negative increments which if applied to any set of prices will cause the demand for all goods in the market to decrease. We shall be using Postulate 4 again in the next section as the basis for our discussion of linear demand functions.

Theorem 5. If the functions $f: R_n^+ \to R_n$ satisfies Postulate 4, then any solution to Problem 3, given $p, s \in R_n^+$, is unique.

Proof. Let p^1 and p^2 be solutions to Problem 3. Without loss of generality we may assume that

$$(p_1^2 - p_1^1)/q_1 \geq |p_i^2 - p_i^1|/q_i, \quad \text{for} \quad i = 2, 3, \ldots, n.$$

If $p_1^2 = p_1^1$ then $p^1 = p^2$ and there is nothing to prove. If $p_1^2 > p_1^1$ either $f_1(p^2) = s_1$ and $f_1(p^1 \leq s_1$ or $f_1(p^1) = 0$ and $f_1(p^2) \geq 0$ so that $f_1(p^2) - f_1(p^1) \geq 0$. Let $\psi: [0, 1] \to R$ be given by

$$\psi(x) = f_1(p^1 + x(p^2 - p^1)).$$

3. The reader should not be surprised if this proof seems familiar. It is essentially an adaptation of McKenzie's proof of international trade equilibrium. See McKenzie (1954) and Dorfman, Samuelson, and Solow (1958, pp. 366–75).

Hence

$$f_1(p^2) - f_1(p^1) = \int_0^1 \psi'(x)dx$$

$$= \int_0^1 \left(\sum_{j=1}^{n} \frac{\partial}{\partial p_j} f_1(p_j^2 - p_j^1)\right) dx$$

$$= \int \left[\frac{\partial f_1}{\partial p_1} (p_1^2 - p_1^1) + \sum_{j=2}^{n} \frac{\partial f_1}{\partial p_j} (p_j^2 - p_j^1)\right] dx$$

$$\leq \int \left[\frac{\partial f_1}{\partial p_1} (p_1^2 - p_1^1) \frac{q_1}{q_1} + \sum_{2}^{n} \frac{\partial f_1}{\partial p_j} (p_1^2 - p_1^1) \frac{q_j}{q_1}\right] dx$$

$$= (p_1^2 - p_1^1)/q_1 \int \left[\sum_{1}^{n} \frac{\partial f_1}{\partial p_j} q_j\right] dx < 0;$$

contradicting $f_1(p^2) - f_1(p^1) \geq 0$. This proves the theorem.

Linear Demand Functions

In this section we shall investigate the implications of some of our postulates for the parameters of a linear function f; that is, a function of the form

$$f_i(p) = a_{i0} + \sum_{1}^{n} a_{ij} p_j, \quad \text{for} \quad i = 1, \ldots, n.$$

Postulates 3 and 4 imply that the coefficients a_{ij} have the following properties.

(1) $a_{i0} > 0$, for $i = 1, \ldots, n$

(2) $a_{ii} < 0$ and $a_{ij} \geq 0$, for $i \neq j$, $i, j = 1, \ldots, n$

(3) There exists a vector $\{q_j\} > 0$ such that $\Sigma a_{ij} q_j < 0$ for $i = 1, \ldots, n$.

Consider the matrix $[b_{ij}] = [(1/q_i a_{ii}) a_{ij} q_j]$. This matrix is unity on and negative off the diagonal, and by Assumption 3 has a positive sum across all rows. Hence $\{b_{ij}\}$ is a Leontief matrix (c.f., Dorfman, Samuelson, and Solow, 1958, pp. 255–56), and has an inverse, $[b]^{-1}$ with all elements positive. We may now state

Theorem 6. *$\{a_{ij}\}$ has an inverse with all elements negative.*

Proof. $\{a_{ij}\} = \{a_{ii}\} Q \{b_{ij}\} Q^{-1}$ where

$$Q = \begin{bmatrix} q_1 & 0 \cdots & 0 \\ 0 & q_2 \cdots & 0 \\ 0 & 0 \cdots & q_n \end{bmatrix}$$

and

$$a_{ii} = \begin{bmatrix} a_{11} & 0 & \cdots & 0 \\ 0 & a_{22} & \cdots & 0 \\ \cdot & \cdot & & \cdot \\ \cdot & \cdot & & \cdot \\ \cdot & \cdot & & \cdot \\ 0 & 0 & & a_{nn} \end{bmatrix}$$

Thus $\{a_{ij}\}$ is a product of nonsingular matrices and has inverse $Q\{b_{ij}\}^{-1}Q^{-1}\{a_{ii}\}^{-1}$. Since $Q > 0$, $Q^{-1} > 0$, and $\{a_{ii}\}^{-1} < 0$; $\{a_{ij}\}^{-1} < 0$.

Corollary. *There exists a set of prices $\{V_j\} > 0$ such that $a_{j0} + \Sigma a_{ij} V_j = 0$.*

Proof. By the nonsingularity of $\{a_{ij}\}$, $V_j = -\Sigma_i a_{ij}^{(-1)} a_{i0} > 0$.

Corollary. *f satisfies Postulates 2 and 3(a). We shall in the sequel consider the V_j's as basic units of value for the commodities in the market. We motivate this terminology by the property of the V_j's that they are the set of least upper bounds of prices of the respective goods at which positive quantities can be sold in the market. Further, under the assumption of symmetry of the demand matrix the set of prices that maximizes total industry revenue is precisely $\{V_j/2\}$.*

For convenience we redefine the unit of measurement for goods as one dollar's worth in terms of the V_j's. This gives us quantity and price variables respectively.

$$\delta_i = V_i d_i, \quad \text{for} \quad i = 1, \ldots, n, \tag{A.1}$$

and

$$q_j = p_j / V_j, \quad \text{for} \quad j = 1, \ldots, n.$$

Letting $\beta = \Sigma V_i a_{i0}$, $w_i = V_i a_{i0}/\beta$, and $\lambda_{ij} = a_{ij} V_j / w_j$, the demand functions in our new units become

$$\delta_i = \beta w_i \left[1 - q_i - \sum_j \lambda_{ij} w_j (q_i - q_j) \right], \quad \text{for} \quad i = 1, \ldots, n. \tag{A.2}$$

In this parameterization, we have β representing the size of the market in natural dollar's worths, the w_j's giving the basic weights of products in this market and the λ_{ij}'s the weight-compensated intergood price dependence.

A restriction on the allowable matrices $\{\lambda_{ij}\}$ can be derived from the Hicks-Slutsky equations for the partial derivatives of individual demand

$$\frac{\partial q_i}{\partial q_j} = \xi_{ij} - \delta_j \frac{\partial q_i}{\partial y} = \beta w_i \lambda_{ij} q_j, \tag{A.3}$$

where ξ_{ij} is the income-compensated partial derivative δ_i with respect to q_j and y is income. Since it is well known (Samuelson, 1945) that $\xi_{ij} = \xi_{ji}$, and

$$\frac{\partial \delta_i}{\partial g_j} - \frac{\partial \delta_j}{\partial g_i} = \delta_i \frac{\partial \delta_j}{\partial y} - \delta_j \frac{\partial \delta_i}{\partial y} = \frac{\delta_i \delta_j}{y}\left[\frac{y}{\delta_j}\frac{\partial \delta_j}{y} - \frac{y}{\delta_i}\frac{\partial \delta_i}{\partial y}\right].$$

If we define η_i as the income elasticity of good i when all prices $q_i = q$, we get from Eqs. (A.2) and (A.3)

$$\beta w_i w_j(\lambda_{ij} - \lambda_{ji}) = \frac{w_i w_j}{y}\beta(1 - q)^2(\eta_j - \eta_i)$$

or

$$\lambda_{ij} - \lambda_{ji} = \frac{\beta(1 - q)^2}{y}(\eta_j - \eta_i). \tag{A.4}$$

Theorem 7. Thus we have proved that a necessary and sufficient condition for the symmetry of $\{\lambda_{ij}\}$ is that products i and j have equal income elasticities when all prices are equal in terms of the V_j's.

Computation

It is of interest if one expects to make economic predictions to be able to compute solutions to Problem 3, especially if one is interested in simulation or gaming. We shall show in this section that under the general conditions of Postulates 1–3 and a slightly weakened version of Postulate 4 for the demand function an algorithm exists for computing these solutions. It is of some disappointment that this algorithm is merely convergent and not finite. However, in the case of a linear demand function which satisfies Postulates 1–4 we can demonstrate another algorithm that is finite.

Let us take up the general case first and pronounce our weakened conditions.

Postulate 4′. $f_i(p)$ is a strictly decreasing function of p_i and a nonincreasing function of p_j where j is not equal to i.

We begin by defining

$$\phi_i(p, y) = f_i(p_i, \ldots, p_{i-1}, y, p_{i+1}, \ldots, p_n)$$

and

$$R_i(p, x) = \{y \mid y \geq 0,\ \phi_i(p, y) = x\}$$

and

$$\psi_i(p, x) = \begin{cases} R_i(p, x) & \text{if } R_i(p, x) \neq \phi, \\ \{0\} & \text{if } R_i(p, x) = \phi. \end{cases}$$

We continue by stating

Lemma 1. $\psi_i(p, x)$ *is a single-valued continuous function on* R^+_{n+1} *nonincreasing in x and nondecreasing in the components of p.*

Proof. Since $\phi_i(p, y)$ is strictly monotone in y, R_i has at most one and ψ_i has exactly one element. Further by Postulate 3, $\phi_i(p, \psi_i(p, x)) \leq x$.

Suppose $\psi_i(p - q, x + \Delta) > \psi_i(p, x) \geq 0$ with $q \geq 0$, $x \geq 0$. Then $\phi_i(p - q, \psi_i(p - q, x + \Delta)) = x + \Delta \geq x$ and $\phi_i(p, \psi_i(p, x)) \leq x$, contradicting Postulate 4′.

It remains to show that ψ_i is continuous. Suppose it is not. Then there is a point (p^0, x^0) in R^+_{n+1} and a sequence (p^k, x^k) with limit (p^0, x^0) such that $\phi_i(p^0, x^0)$ is not the limit of $\phi_i(p^k, x^k)$. We consider two distinct cases

(1) $\phi_i(p^0, \psi_i(p^0, x^0)) < x^0$

(2) $\phi_i(p^0, \psi_i(p^0, x^0)) = x^0$.

In case (1), $\psi_i(p^0, x^0) = 0$. By the continuity of f (and hence ϕ_i), there is a neighborhood L of (p^0, x^0), such that if $(p, x) \in L$, then $\phi_i(p, 0) \leq x$, implying $\psi_i(p, x) = 0$. This contradicts the hypothesis of discontinuity at (p^0, x^0).

In case (2), suppose that (p^k, x^k) has no subsequence such that $\psi_i(p^k, x^k) > 0$. Then, if $\psi_i(p^0, x^0) > 0$, and $\phi_i(p^0, 0) > x^0$ and $\phi_i(p^k, 0) \leq x^k$ contradicting the continuity of ϕ_i. If $\psi_i(p^0, x^0) > 0$, the discontinuity hypothesis is contradicted.

We are left with the case that $\psi_i(p^k, x^k) > 0$. Then for some $\epsilon > 0$ and some subsequence (p^l, x^l) either

(a) $\psi_i(p^l, x^l) - \psi_i(p^0, x^0) > \epsilon$, or

(b) $\psi_i(p^l, x^l) - \psi_i(p^0, x^0) < -\epsilon$.

If (a) holds $x^l = \phi_i(p^l, \psi_i(p^l, x^l)) < \phi_i(p^l, \psi_i(p^0, x^0) + \epsilon$. The left-hand side of this inequality has x^0 as limit; and the right-hand has as limit a number strictly less than x^0. The contradiction is similarly shown if (b) holds. This completes the proof of the lemma.

We now define an algorithm for solving Problem 3.

Algorithm I. Let $p, q \geq 0$ *be given.*

$$p_i^1 = 0,$$

and

$$p_i^{t+1} = \min(\psi_i(p^t, 0), \max[\psi_i(p^t, q_i), p_i)], \quad \text{for} \quad i = 1, 2, \ldots, n.$$

Theorem 8. The sequence (p^t) *in Algorithm I increases monotonically with limit* p'*, the solution to Problem 3.*

Proof. Since $\psi_i(p^t, x)$, p_i are monotonically increasing functions of p^t and the min and max of increasing functions are increasing functions, we can write $p^{t+1} = K(p^t)$ where K is an increasing function. $p^{t+1} - p^t =$

$K(p^t) - K(p^{t-1})$ and if $p^t \geq p^{t-1}$ then $p^{t+1} \geq p^t$. Since $K(p^t) \geq 0$ and $p' = 0$, p^t is an increasing sequence.

We assert that $f(p^t) \geq 0$. First, by Postulate 3(a), $f(p') \geq 0$. Suppose that $f(p^k) \not\geq 0$ and $f(p^j) \geq 0$ for $j < k$. For some i, $f_i(p^k)$, $\phi_i(p^k, p_i^k) < 0$. By monotonicity $\phi_i(p^k, p_i^k) \geq \phi_i(p^{k-1}, p_i^k)$. Hence $p_i^k > \psi_i(p^{k-1}, 0)$, which contradicts the definition of the algorithm. Since $f(p^t) \geq 0$, by Postulate 2, (p^t) is bounded and has a limit p'. Since K is continuous $p' = K(p')$. We assert that p' solves Problem 3.

For each i, since by Lemma 1, $\psi_i(p', 0) \geq \psi_i(p', q_i)$ either:

(1) $p_i > p_i' = \psi_i(p', 0)$, or

(2) $p_i < p_i' = \psi_i(p', q_i)$, or

(3) $\psi_i(p', q_i) \leq p_i = p_i \leq \psi_i(p', 0)$.

If (1) holds, $f_i(p') \leq 0$, hence $f_i(p') = 0$. If (2) holds, $f_i(p') \leq q_i$, but $f_i(p') \neq q_i$ implies $p_i' = 0$ contradicting $p_i' > p_i$. Finally if (3) holds, $0 \leq f_i(p') = \phi_i(p', p_i) \leq \phi_i(p', \psi_i(p', q_i)) \leq q_i$.

We have shown that p' has the precise properties to be a solution of Problem 3 and the theorem is proved.

The only issue remaining for the practical application of this algorithm is the computation of the functions ψ_i. In some cases it may be possible to get simple closed form expressions for them, but in general it must be expected that some iterative calculation such as Newton's method may be required. It may be possible in some cases to derive an algorithm from Algorithm I which is usually finite. A solution to Problem 3 determines a threefold partition of $\Pi = (S_+, S_-, S_0)$, of the set of goods $S = \{1, 2, \ldots, n\}$ determined by $S_+ = \{i | p_i' > p_i\}$, $S_- = \{i | p_i' < p_i\}$, and $S_0 = \{i | p_i' = p_i\}$.

If we start with such a partition, we would get a set of simultaneous equations,

$$\begin{aligned} i \in S_+ &\quad \text{implies} \quad f_i(p') = q_i, \\ i \in S_- &\quad \text{implies} \quad f_i(p') = 0, \\ i \in S_0 &\quad \text{implies} \quad p_i' = p_i. \end{aligned}$$

If this set of n equations in n variables, can be solved then, to every partition there corresponds a price vector p'. The question of whether a particular partition of the 3^n possible partitions yields a solution is precisely the question of whether the inequalities:

$$\begin{aligned} i \in S_+ &\quad \text{implies} \quad p_i' > p_i, \\ i \in S_- &\quad \text{implies} \quad p_i' < p_i, \\ i \in S_0 &\quad \text{implies} \quad 0 \leq f_i(p') \leq q_i, \end{aligned}$$

are satisfied.

It is apparent that Algorithm I determines a sequence of partitions (Π^t)

which, since p^t increases monotonically, does not cycle. It is also apparent that, unless for some i, p_i^t converges to p_i without reaching it finitely, the partition determined by the solution will be reached in a finite number of steps.

Whether this algorithm will be a useful computational scheme will depend on whether there exists an economical method of calculating the solution to the above simultaneous equations. It is difficult to characterize that class of functions f which will make such a calculation appealing.

Before we turn to our finite algorithm for the linear case it will be of some interest to inject a remark concerning a connection between Problem 3 and constrained maximization problems.

Remark. If f^{-1} exists and is the gradient of some function F, and if y is a solution to: max $F - py$ subject to $0 \leq y \leq q$, then $p' = f^{-1}(y)$ is a solution to Problem 3.

This follows directly from the Kuhn-Tucker conditions for constrained maxima which in this case are equivalent to:

$$\begin{aligned} 0 < y_i < q_i \quad &\text{implies} \quad f_i^{-1}(y) - p_i = 0, \\ y_i = 0 \quad &\text{implies} \quad f_i^{-1}(y) - p_i < 0, \\ y_i = q_i \quad &\text{implies} \quad f_i^{-1}(y) - p_i > 0. \end{aligned}$$

Since $p_i' = f_i^{-1}(y)$ and $y_i = f_i(p')$, this is the same as the statement of Problem 3. In the linear case $f(p) = a_0 + A_p$ would satisfy the hypothesis of the remark if A were symmetric. In Theorem 7 we showed that the symmetry of A corresponds to equal-income elasticities of the goods. This is not a very unreasonable restriction for an oligopoly market model. When we consider that Postulate 4 implies that $-A$ is a scale transform of a Leontief matrix which is known to be positive definite (cf., Dorfman, Samuelson, and Solow, 1958, p. 256, condition 9A), we see that the associated quadratic programming problem is solvable by the standard algorithms.

However we shall deal with the linear case of Problem 3 with a special algorithm that exploits the Leontief "diagonal dominance" property we have postulated for the matrix A.

For convenience we shall rescale the parameters of Problem 3 in the following way. Let $x = p' - p$, $u = (a_0 + Ap)$, and $v = q + u$. We shall also assume that the variables are scaled so that the row sums of A are negative. This will not be necessary for the calculation of the algorithm but will facilitate the proofs. Problem 3 can now be restated as

Problem 3A. Given vectors $v \geq u$ and $-A$, a Leontief matrix with row sums positive, find a vector x such that

(1) $u \leq Ax = y \leq v$,

(2) $x_i > 0$ implies that $y_i = v_i$,
(3) $x_i < 0$ implies that $y_i = u_i$.

Any solution to Problem 3A will be unique by Theorem 5. As in the previous discussion we will consider partitions of the form (S_+, S_-, S_0) such that for $i \in S_+$, $y_i = v_i$, for $i \in S_-$, $y_i = u_i$, and for $i \in S_0$, $x_i = 0$. We will consider the variables and matrix A to be partitioned so that we have

$$\begin{bmatrix} y^+ \\ y_- \\ y_0 \end{bmatrix} = \begin{bmatrix} A_{++} & A_{+-} & A_{+0} \\ A_{-+} & A_{--} & A_{-0} \\ A_{0+} & A_{0-} & A_{00} \end{bmatrix} \begin{bmatrix} x_x \\ x_- \\ x_0 \end{bmatrix}.$$

Since a diagonal submatrix of a Leontief matrix is also a Leontief matrix, it has a non-negative inverse and we can express the pair (x, y) in terms of the basis (x_0, y_+, y_-).

$$(1)\quad \begin{bmatrix} x_+ \\ x_- \end{bmatrix} = \begin{bmatrix} A_{++} & A_{+-} \\ A_{-+} & A_{--} \end{bmatrix}^{-1} \left(\begin{bmatrix} y_+ \\ y_- \end{bmatrix} - \begin{bmatrix} A_{+0} \\ A_{-0} \end{bmatrix} x_0 \right)$$

$$(2)\quad y_0 = [A_{0+} \quad A_{0-}] \begin{bmatrix} A_{++} & A_{+-} \\ A_{-+} & A_{--} \end{bmatrix}^{-1} \begin{bmatrix} y_+ \\ y_- \end{bmatrix}$$
$$+ \left(A_{00} - [A_{0+} \quad A_{0-}] \begin{bmatrix} A_{++} & A_{+-} \\ A_{-+} & A_{--} \end{bmatrix}^{-1} \begin{bmatrix} A_{+0} \\ A_{-0} \end{bmatrix} \right) x_0.$$

Lemma 2. If

$$B = \begin{bmatrix} B_{11} & B_{12} \\ B_{21} & B_{22} \end{bmatrix}$$

is a diagonal partition of a Leontief matrix, then $B' = (B_{11} - B_{12}B_{22}^{-1}B_{21})$ *is also a Leontief matrix.*

Proof. Since B_{12}, $B_{21} \leq 0$ and $B_{22}^{-1} \geq 0$, B' must be nonpositive off the diagonal since B_{11} is. Letting f and g be vectors whose components are equal to one, we have by hypothesis $B_{11}f + B_{12}g \geq 0$, and $B_{21}f + B_{22}g \geq 0$. Since $B_{22}^{-1} \geq 0$ and $B_{21} \leq 0$, we have $B_{22}^{-1}B_{21}f + g \geq 0$ or $g \geq -B_{22}^{-1}B_{21}f \geq 0$. Hence since $B_{12} \leq 0$ we have $0 \leq B_{11}f + B_{12}g \leq B_{11}f - B_{12}B_{22}^{-1}B_{12}f = (B_{11} - B_{12}B_{22}^{-1}B_{21})f$.

We now pose a special problem which is a subproblem of Problem 3A.

Problem 4. Let S_+ be given. Find a vector x such that $y_i = v_i$ when $i \in S_+$, and then $i \notin S_+$, $x_i \leq 0$, $y_i \geq u_i$ and $x_i < 0$ imples $y_i = u_i$.

By Lemma 2, Problem 4 can be translated into a version of Problem 3 with Postulate 4 satisfied. Hence any solution is unique.

Lemma 3. Let S_+ and $\bar{x}$ the solution to Problem 4 be given. If $\hat{x}$ is a vector such that $\hat{y}_i = v_i$ when $i \in S_+$ and, when $i \notin S_+$, $\hat{x}_i \leq 0$ and $\hat{y}_i \geq u_i$, then $\hat{x} \leq \bar{x}$.

Proof. Consider the partition (S_+, ϕ, S_0) and the corresponding basis representation of y_0. Let

$$y_0 = A_{0+}A_{++}^{-1}v_+ + (A_{00} - A_{0+}A_{++}^{-1}A_{+0})x_0$$

be written as $y_0 = \beta + \alpha z$. We define for $i \in S_0$, $\phi_i(z) = -(\beta_i - u_i + \Sigma_{j\neq i}\alpha_{ij}z_j)/\alpha_{ij}$. Let the sequence (z^t) be defined by

$$z^1 = \hat{x}_0$$

and

$$z_i^{t+1} = \min(0, \psi_i(z^t)).$$

First, we show that (z^t) is an increasing sequence. In the proof of Theorem 8 we have essentially proved that if $z^t \geq z^{t-1}$ then $z^{t+1} \geq z^t$, and $\hat{y}_0 \geq u_0$ implies $z_1^1 \leq \psi_i(z^1)$ or $z^2 \geq z^1$. Since (z^t) is bounded above by 0 it has a limit z^0. It remains to show that z^0 is a solution to Problem 4. By continuity of ψ_i, $z_i^0 = \min(0, \psi_i(z^0))$. Hence for each $i \in S_0$, either $z_i^0 = 0$ and $y_i \geq u_i$ or $z_i^0 < 0$ and $y_i = u_i$.

We shall now define a finite algorithm for solving Problem 4 which will be part of the algorithm for solving Problem 3A.

Algorithm II. We are given a partition, (S_+, S_-, S_0) and a corresponding basis representation with $y_+ = u_+$, $y_- = u_-$, and $x_0 = 0$. We distinguish three cases:

(1) $x_- \nleq 0$,
(2) $x_- \leq 0$, $y_0 \ngeq u_0$,
(3) $x_- \leq 0$, $y_0 \geq u_0$.

In case (1), the algorithm step is as follows: Fix $i \in S_-$ such that $x_i > 0$. Increase y_i to the value that makes $x_i = 0$. Transfer i from S_- to S_0.

In case (2), fix $i \in S_0$ such that $y_i < u_i$. Decrease x_i to the value that makes $y_i = u_i$ and transfer from S_0 to S_-.

In case (3), the algorithm is terminated.

Theorem 9. Algorithm II finitely computes the solution to Problem 4. Obviously, case (1) must terminate by exhaustion. Case (2) must also terminate if case (1) does not recur. Since in each step of case (2) a component of x_0 decreases, by Lemma 2 x_- cannot increase and case (1) cannot recur. It is obvious that case (3) is a solution to Problem 4.

We are finally ready to define our algorithm for Problem 3A.

Algorithm III. *Step 1: Let $S_+ = S_- = \Theta$, and $S_0 = S$. Apply Algorithm II. Step t: Where $t > 1$: We are given a partition (S_+, S_-, S_0) and a corresponding basis representation with $y_+ = v_+$, $y_- = u_-$ and $x_0 = 0$. We distinguish two cases:*

(1) $y_0 \not\leq v_0$

(2) $y_0 \leq v_0$.

In case (1), the algorithm step is to fix $i \in S_0$ such that $y_i > v_i$ and increase x_i to the value that makes $y_i = v_i$, then transfer i from S_0 to S_+. Now apply Algorithm II.

In case (2), the algorithm is terminated.

Theorem 10. *Algorithm III finitely computes the solution to Problem 3A.*

Proof. Obviously the algorithm terminates finitely since each step increases S_+ by one element. To show that case (2) is a solution we must show that $x_- \leq 0$, $x_+ \geq 0$ and $u_0 \leq y_0 \leq v_0$. Now Algorithm II and the definition of case (2) give all but $x_+ \geq 0$.

We shall show that $x_+ \geq 0$ by induction on the algorithm steps. At the end of step 1 $x_+ \geq 0$ is vacuously true. Let us assume $x_+ \geq 0$ at the end of step $t - 1$. In step t some component i is transferred to S_+ and Problem 4 is solved. Let S_+^{t-1} be S_+ at the end of step $t - 1$. Consider the partition $\{S_+^{t-1}, \Theta, S_0\}$, where $y_+ = y_+^{t-1}$, $x_0 = (x_-^{t-1}, x_0^{t-1})$. Increase x_i to the value that makes $y_i = v_i$ and transfer i from S_0 to S_+. Let $\hat{x}$ equal the transformed value of x. The transformation will cause an increase in x_+ and for all i not in S_+^t, $x_i \leq 0$. Further all components of y_0 will have increased so for all i not in S_+^t, $y_i > u_i$. We have $\hat{x} \geq x^{t-1}$ and by Lemma 3, $x^t \geq \hat{x}$, *since solutions of Problem 4 are unique.*

It is to be noted that the simplex tableau is very useful for this algorithm. The calculation proceeds by a series of Gaussian eliminations whose number rarely exceeds the number of goods in the problem.

Bibliography

AAKER, D. A. 1971. The new-trier stochastic model of brand choice. *Management Science 17:B435*–450.

ADELMAN, M. A. 1969. Comment on the "H" concentration measure as a numbers-equivalent. *Review of Economics and Statistics* 51:99–101.

AKERLOF, G. A. 1970. The market for "lemons": qualitative uncertainty and the market mechanism. *Quarterly Journal of Economics* 84:488–500.

AMERICAN INSTITUTE OF CERTIFIED PUBLIC ACCOUNTANTS. 1971. *APB accounting principles.* New York.

ARROW, K. J. 1974. *Limits of organization.* New York: W. W. Norton.

———, AND G. DEBREU. 1954. Existence of equilibrium for a competitive economy. *Econometrica* 22:265–290.

ARROW, K. J., AND F. H. HAHN. 1971. *General competitive analysis.* San Francisco: Holden-Day.

AUMANN, R. J. 1973. Disadvantageous monopolies. *Journal of Economic Theory* 6:1–11.

BAIN, J. S. 1956. *Barriers to new competition.* Cambridge, Mass.: Harvard University Press.

BALDERSTON, F. E., AND A. C. HOGGATT. 1962. Simulation of market processes Berkeley, Institute of Business and Economic Research, Special Publication #1.

BARON, D. 1970. Price uncertainty, utility and industry equilibrium in pure competition. *International Economic Review* 11:463–480.

BASS, F. M., R. D. BUZZELL, GREENE W. LAZER, E. A. PESSEMIER, D. L. SHAWVER, A. SHUCHMAN, C. THEODORE, AND G. W. WILSON. 1961. *Mathematical models and methods in marketing.* Homewood, Ill.: Irwin.

BAUMOL, W. J. 1959. *Business behavior, value and growth.* New York: Macmillan.

BECKMANN, M. J. 1965. Edgeworth-Bertrand duopoly revisited. In *Operations research-verfahren III,* ed. Rudolf Henn. Meisenheim: Sonderdruck Verlag.

______. 1968. *Location theory.* New York: Random House.

BERTRAND, J. 1883. Théorie mathématique de la richesse sociale (Review). *Journal des Savants* 68:499–508.

BETTMAN, J. R. 1971. A graph theory approach to comparing consumer information processing models. *Management Science* 18:P-114–P-128.

BISHOP, R. L. 1952. Elasticities and market relationship. *American Economic Review* 42:779–803.

BÖHM-BAWERK, E. VON. 1923. *Positive theory of capital,* translated by W. Smark. 1891; rpt. New York: Steckert.

BONINI, C. P. 1963. *Simulation of information and decision systems in the firm.* Englewood Cliffs, N.J.: Prentice-Hall.

BORCH, K. 1968. *The economics of uncertainty.* Princeton, N.J.: Princeton University Press.

BOULDING, K. E. 1962. *A reconstruction of economics.* New York: Science Editions.

BOWLEY, A. L. 1928. Bilateral monopoly. *Economic Journal* 38:651.

BOWMAN, W. S. 1973. *Patents and antitrust law: A legal and economic appraisal.* Chicago: University of Chicago Press.

BREMS, H. 1951. *Product equilibrium under monopolistic competition.* Cambridge, Mass.: Harvard University Press.

BRONFENBRENNER, M. 1940. Applications of the discontinuous oligopoly demand curve. *Journal of Political Economy* 48:420–427.

BUCHAN, J., AND E. KOENIGSBERG. 1963. *Scientific inventory management.* Englewood Cliffs, N.J.: Prentice-Hall.

CHAMBERLIN, E. H. 1962. *The theory of monopolistic competition,* 8th ed. 1933; rpt. Cambridge, Mass.: Harvard University Press.

CHARNES, A., AND W. W. COOPER. 1961. *Management models and industrial applications of linear programming.* New York: Wiley.

CLARKSON, G. P. E. 1962. *Portfolio selections: a simulation of trust investment.* Englewood Cliffs, N.J.: Prentice-Hall.

COHEN, K. J. 1960. *Computer models of the shoe, leather, hide sequence.* Englewood Cliffs, N.J.: Prentice-Hall.

______, W. R. DILL, A. A. KUEHN, AND P. R. WINTERS. 1964. *The Carnegie Tech management game.* Homewood, Ill.: Irwin.

COURNOT, A. A. 1897. *Recherches sur les principes math*ématiques de la théorie *des richesses,* translated by N. T. Bacon. 1838; rpt. New York: Macmillan.

CYERT, R., AND J. MARCH. 1955. Organizational structure and pricing behavior in an oligopolistic market. *American Economic Review* 45:129–139.

______. 1963. *A behavioral theory of the firm.* Englewood Cliffs, N.J.: Prentice-Hall.

DEBREU, G. 1959. *Theory of value.* New York: Wiley.

______. 1975. The rate of convergence of the core. *Journal of Mathematical Economics* 2:1–7.

———, AND H. SCARF. 1963. A limit theorem on the core of an economy. *International Economic Review* 4:235–246.

DORFMAN, R., AND P. O. STEINER. 1957. Optimal advertising and optimal quality. *American Economic Review* 47:826–836.

DORFMAN, R., P. A. SAMUELSON, AND R. M. SOLOW. 1958. *Linear programming and economic analysis.* New York: McGraw-Hill.

DUBEY, P., AND M. SHUBIK. 1977a. A closed economic system with production and exchange modelled as a game of strategy. *Journal of Mathematical Economics* 4:253–287.

———. 1977b. Trade and prices in a closed economy with exogenous uncertainty. *Econometrica* 45:1657–1680.

———. 1978a. The noncooperative equilibria of a closed trading economy with market supply and bidding strategies. *Journal of Economic Theory* 17:1–20.

———. 1978b. On the foundations of the theory of monopolistic competition. Yale University, Cowles Foundation Discussion Paper #484.

———. 1978c. The profit maximizing firm: managers and stockholders. Yale University, Cowles Foundation Discussion Paper #483.

———. 1980. Information conditions, communication and general equilibrium. *Mathematics of Operations Research,* forthcoming.

EDGEWORTH, F. Y. 1881. *Mathematical psychics.* London: Kegan Paul.

———. 1925. *Papers relating to political economy,* vol. I. London: Macmillan.

FELLNER, W. 1949. *Competition among the few.* New York: Alfred Knopf.

FIRESTONE, O. J. 1967. *The economic implications of advertising.* Toronto: Methuen.

FISHBURN, P. C. 1974. Lexicographic orders, utilities and decision rules: a survey. *Management Science* 20:1442–1471.

FITZROY, F. 1974. Monopolistic equilibrium, non-convexity and inverse demand. *Journal of Economic Theory* 7:1–16.

FRIEDMAN, J. W. 1977. *Oligopoly and the theory of games.* Amsterdam: Elsevier North-Holland.

FRIEDMAN, L. 1956. A competitive bidding strategy. *Operations Research* 4:104–112.

GABOR, A., AND C. W. GRANGER. 1966. Price as an indicator of quality: report on enquiry. *Economica* 33:43–70.

GABSZEWICZ, J. J., AND J. P. VIAL. 1972. Oligopoly "à la Cournot" in general equilibrium analysis. *Journal of Economic Theory* 4:381–400.

GALE, D. 1960. *The theory of linear economic models.* New York: McGraw-Hill.

GASKINS, D. W., JR. 1971. Dynamic limit pricing, optimal pricing under threat of entry. *Journal of Economic Theory* 3:306–322.

GREISMER, J. H., and M. SHUBIK. 1963. Towards a study of bidding processes, part 2: games with capacity limitations. *Naval Research Logistics Quarterly* 10:151–173.

GUPTA, S. K., and K. S. KRISHNAN. 1967. Mathematical models in marketing. *Operations Research* 15:1040–1050.

HARROD, R. F. 1951. *Economic essays.* London: Macmillan.

HARSANYI, J. D. 1963. A simplified bargaining model for the *n*-person cooperative game. *International Economic Review* 4:194–220.

——— 1967a. Games with incomplete information played by "Bayesian" players, Part I. *Management Science* 13:159–182.

———. 1967b. Games with incomplete information played by "Bayesian" players, Part II. *Management Science* 14:320–324.

———. 1968. Games with incomplete information played by "Bayesian" players, Part III. *Management Science* 15:486–502.

———. 1975. The tracing procedure: a Bayesian approach to defining a solution for *n*-person noncooperative games. *International Journal of Game Theory* 4:61–94.

HENDERSON, J. M., AND R. QUANT. 1971. *Microeconomic theory*, 2nd ed. 1958; rpt. New York: McGraw-Hill.

HENSHAW, R. C., AND J. R. JACKSON. 1966. *The executive game.* Homewood, Ill.: Irwin.

HICKS, J. R. 1939. *Value and capital.* London: Oxford University Press.

———. 1954. The process of imperfect competition. *Oxford Economic Papers,* 6:41–54.

HOTELLING, H. 1929. Stability in competition. *Economic Journal* 39:47–51; rpt. in *Readings in price theory.* 1952. K. E. Boulding and G. J. Stigler, eds. Homewood, Ill.: Irwin.

IJIRI, J., AND H. A. SIMON. 1964. Business firm growth and size. *American Economic Review* 54:77–89.

JONES, J. M. 1970. A dual-effects model of brand choice. *Journal of Marketing Research* 7:458–464.

JOSKOW, P. 1973. Pricing decisions of regulated firms: a behavioral approach. *Bell Journal of Economics* 4:118–140.

KAMIEN, M. I., and N. L. SCHWARTZ. 1975. Cournot oligopoly with uncertain entry. *Review of Economic Studies* 42:125–131.

KINBERG, Y., A. G. RAO, AND M. F. SHAKUN. 1974. A mathematical model for price promotions. *Management Science* 20:948–959.

KIRMAN, A. P., AND M. J. SOBEL. 1974. Dynamic oligopoly with inventories. *Econometrica* 42:279–287.

KRELLE, W. 1961. *Preistheorie.* Tübingen: Mohr Siebeck.

———. 1968. *Praferenz und entscheidungstheorie.* Tübingen: Mohr.

LAFFONT, J.-J., AND G. LAROQUE. 1976. Existence d'un équilibre général de concurrence imparfaite: une introduction. *Econometrica* 20:283–294.

LEITCH, R. A. 1974. Marketing strategy and the optimal production schedule. *Management Science* 21:302–312.

LELAND, H. E. 1972a. The dynamics of a revenue maximizing firm. *International Economic Review* 13:376–385.

———. 1972b. The firm and uncertain demand. *The American Economic Review* 62:278–291.

LERNER, A. P. 1934. The concept of monopoly and the measurement of monopoly power. *Review of Economic Studies* 1:157–175.

LEVITAN, R. E. 1964. Oligopoly demand. Yorktown Heights, N.Y.: IBM Watson Research Center, RC-1239.

———, AND M. SHUBIK. 1961. *The financial, allocation and marketing executive game.* Sands Point, N.Y.: IBM Executive Dept.

———. 1971. Price variation duopoly with differentiated products and random demand. *Journal of Economic Theory* 3:23–39.

———. 1972. Price duopoly and capacity constraints. *International Economic Review* 13:111–122.

———. 1978. Duopoly with price and quantity as strategic variables. *International Journal of Game Theory* 6:1–4.

LIFF, J. 1969. Purchasing policies for seasonal style goods, case study and analysis. Ph.D. dissertation, Yale University, Department of Administrative Sciences.

LOVELL, M. C. 1970. Product differentiation and market structure. *Western Economic Journal* 8:120–143.

LUCE, R. D., AND H. RAIFFA. 1957. *Games and decisions*. New York: Wiley.

MCCALL, J. J. 1965. The economics of information and optimal stopping rules. *Journal of Business* 38:300–317.

———. 1970. Economics of information and job search. *Quarterly Journal of Economics* 84:113–126.

MCKENZIE, LIONEL. 1954. On equilibrium in Graham's model of world trade. *Econometrica* 22:147–161.

MACHLUP, F. 1952a. The characteristics and classifications of oligopoly. *Kyklos* 5:145.

———. 1952b. *The economics of sellers competition*. Baltimore: Johns Hopkins.

MANN, H. MICHAEL. 1966. Seller concentration barriers to entry, and rates of return in thirty industries, 1950–1960. *Review of Economics and Statistics* 48:296–307.

MARCH, J. G., and H. A. SIMON. 1958. *Organizations*. New York: Wiley.

MARRIS, R. 1964. *The economic theory of managerial capitalism*. Glencoe, Ill.: Free Press.

MARSCHAK, T., AND R. SELTEN. 1974. *General equilibrium with price-making firms*. Berlin: Springer-Verlag.

MAURIELLO, J. A. 1971. *Accounting for the financial analyst*. Homewood, Ill.: Irwin.

MIYASAWA, K. 1962. An economic survival game. *Journal of Operations Research Society of Japan* 4:94–113.

MONTGOMERY, D. B., ED. 1971. Preface to *Management Science:* Marketing management models. In a special issue of *Management Science* 18:1–2.

———, AND G. L. URBAN. 1969. *Management science in marketing*. Englewood Cliffs, N.J.: Prentice-Hall.

MURRAY, G. R., JR., AND E. A. SILVER. 1966. A Bayesian analysis of the style goods inventory problem. *Management Science* 12:785–797.

NASH, J. F. 1950. Equilibrium points in *N*-person games. *Proceedings of the National Academy of Sciences of the U.S.A.* 36:48–49.

———. 1951. Non-cooperative games. *Annals of Mathematics* 54:286–295.

———. 1953. Two person cooperative games. *Econometrica* 21:128–140.

NEALE, A. D. 1970. *The antitrust laws of the United States*, 2nd ed. London: Cambridge University Press.

NEGISHI, T. 1961. Monopolistic competition and general equilibrium. *Review of Economic Studies* 28:196–201.

NELSON, R. R., AND S. G. WINTER. 1974. Neoclassical versus evolutionary theories of economic growth: critique and prospectus. *The Economic Journal* 84: 886–905.

______. 1978. Forces generating and limiting concentration under Schumpeterian competition. *The Bell Journal of Economics* 9:542–584.

NERLOVE, M., AND K. J. ARROW. 1962. Optimal advertising policy under dynamic conditions. *Economica* 22:129–142.

NIKAIDO, H. 1975. *Monopolistic competition effective demand.* Princeton: Princeton University Press.

NTI, K. O., and M. SHUBIK. 1979. Entry in oligopoly theory: a survey. *Eastern Economic Journal* 5:271–289.

OXENFELDT, A. R., ED. 1963. *Models of markets.* New York: Columbia University Press.

PACKER, H. L. 1963. *The state of research in antitrust law.* New Haven, Conn.: Walter Meyer Research Institute of Law.

PAPANDREOU, A., AND J. T. WHEELER. 1954. *Competition and its regulation.* Englewood Cliffs, N.J.: Prentice-Hall.

PEARCE, F. 1961. An exact method of consumer demand analysis. *Econometrica* 29:499–516.

PERROUX, F. 1975. *Unités* actives et mathématiques nouvelles. Paris: Dunod.

PFOUTS, R. W., AND C. E. FERGUSON. 1959. Market classification systems in theory and policy. *Southern Economic Journal* 26:111–118.

POSNER, R. A. 1969. Oligopoly and the antitrust laws: a suggested approach. *Stanford Law Review* 21:1562–1606.

PRIMEAUX, W., JR., AND M. BOMBALL. 1974. A reexamination of the kinky oligopoly demand curve. *Journal of Political Economy* 82:845–862.

QUALLS, DAVID. 1972. Concentration barriers to entry, and long-run economic profit margins. *Journal of Industrial Economics* 20:146–158.

RADNER, R. 1968. Competitive equilibrium under uncertainty. *Econometrica* 36:31–58.

RAIFFA, H. 1968. *Decision analysis—introducing lectures on choices under uncertainty.* Reading, Mass.: Addison-Wesley.

RAPOPORT, A., and A. M. CHAMMAH. 1965. *Prisoner's dilemma.* Ann Arbor: University of Michigan Press.

RAPOPORT, A., AND M. GUYER. 1966. A taxonomy of 2 × 2 games. *General Systems* 11:203–214.

RHODES, S. A. 1970. Concentration, barriers to entry and rates of return: a note. *Journal of Industrial Economics* 19:82–88.

RICHARDSON, L. F. 1960. *Arms and insecurity.* Chicago: Quadrangle Press; and Pittsburgh: Boxwood Press.

ROBERT MORRIS ASSOCIATES. 1969. *Annual statement studies.*

ROBERTS, J., AND H. SONNENSCHEIN. 1977. On the foundations of the theory of monopolistic competition. *Econometrica* 45:101–113.

ROTHKOPF, M. H. 1969. A model of rational competitive bidding. *Management Science* 15:362–373.

ROTHSCHILD, M. 1973. Models of market organization with imperfect information: a survey. *Journal of Political Economy* 81:1283–1308.

SAMUELSON, P. A. 1945. *The foundations of economic analysis*. Cambridge, Mass.: Harvard University Press.

SCHELLING, T. C. 1960. *The strategy of conflict*. Cambridge, Mass.: Harvard University Press

SCHERER, F. 1970. *Industrial market structure and economic performance*. Chicago: Rand McNally.

SCHRIEBER, A. N., ED. 1970. *Corporate simulation models*. Seattle: University of Washington Press.

SCHUMPETER, J. A. 1950. *Capitalism, socialism and democracy*, 3rd ed. New York: Harper.

SELTEN, R. 1970. *Preispolitik der Mehrproduktenunternehmung in der statischen theorie*. Berlin: Springer-Verlag.

———. 1973. A simple model of imperfect competition where 4 are few and 6 are many. *International Journal of Game Theory* 2:141–201.

———. 1975. Reexamination of the perfectness concept for equilibrium points of extensive games. *International Journal of Game Theory* 5:25–55.

SHAPLEY, L. S. 1975. An example of a slow-converging core. *International Economic Review* 16:345–351.

———. 1976. Noncooperative general exchange. In *Theory and measurement of economic externalities*, S. A. Y. Lin., ed. New York: Academic.

———, AND M. SHUBIK. 1969a. On market games. *Journal of Economic Theory* 1:9–25.

———. 1969b. Price strategy oligopoly with product variation. *Kyklos* 22:30–44.

———. 1971–1974. Game theory in economics. Santa Monica, RAND R-904-NSF.

———. 1972. The assignment game I: the core. *International Journal of Game Theory* 1:111–130.

———. 1974. Game theory in economics. 1972, RAND R-904/2-NSF; also R-904/6-NSF.

———. 1977. Trade using one commodity as a means of payment. *Journal of Political Economy* 85:937–968.

SHERMAN, R. 1972. *Oligopoly: an empirical approach*. Lexington, Md.: D. C. Heath.

SHITOVITZ, B. 1973. Oligopoly in markets with a continuum of traders. *Econometrica* 41:467–502.

SHUBIK, M. 1959a. Edgeworth market games. In *Contributions to the theory of games*, IV. A. W. Tucker and R. D. Luce, eds. Princeton: Princeton University Press.

———. 1959b. *Strategy and market structure*. New York: Wiley.

———. 1961. Objective functions and models of corporate optimization. *Quarterly Journal of Economics* 75:345–375.

———. 1968. A further comparison of some models of duopoly. *Western Economic Journal* 6:260–276.

———. 1971. Games of status. *Behavioral Science* 16:117–129.

———. 1973a. Commodity money, oligopoly, credit and bankruptcy in a general equilibrium model. *Western Economic Journal* 11:24–38.

———. 1973b. Information, duopoly and competitive markets. *Kyklos* 26: 736–761.

———. 1975. *Games for society, business and war*. Amsterdam: Elsevier North-Holland.

———, AND R. E. LEVITAN. 1962. A business game for teaching and research purposes. IBM Research Report RC-731.

———, AND G. L. THOMPSON. 1959. Games of economic survival. *Naval Research Logistics Quarterly* 6:111–123.

SIMON, J. 1969. A further test of the kinky oligopoly demand curve. *American Economic Review* 59:971–975.

SIMON, H. A., AND C. P. BONINI. 1958. The size distribution of business firms. *American Economic Review* 48:607–617.

SMITH, V. L. 1967. Experimental studies of discrimination vs competition in sealed bid auction markets. *Journal of Business* 40:56–82.

SOBEL, M. J. 1971. Noncooperative stochastic games. *Annals of Mathematical Statistics* 42:1930–1935.

———. 1973. Continuous stochastic games. *Journal of Applied Probability* 10:597–604.

SPENCE, M. 1974. Competitive and optimal responses to signals: an analysis of efficiency and distribution. *Journal of Economic Theory* 7:296–332.

STACKELBERG, H. VON. 1934. *Marktform und gleichgewicht*. Berlin: Julius Springer.

———. 1952. *The theory of the market economy*, translation and Introduction by A. T. Peacock. London: William Hodge.

STARK, R. H., AND M. H. ROTHKOPF. 1979. Competitive bidding: a comprehensive bibliography. *Operations Research* 27:484–490.

STATISTICAL ABSTRACT OF THE UNITED STATES. 1976.

STIGLER, G. 1947. The kinky oligopoly demand curve and rigid prices. *Journal of Political Economy* 55:432–449.

SUNDAY TIMES WEEKLY REVIEW. 1969. June 8, p. 45.

SWEEZY, PAUL. 1939. Demand under conditions of oligopoly. *Journal of Political Economy* 47:568–573.

SYLOS-LABINI, PAOLO. 1962. *Oligopoly and technical progress, translated from Italian* by Elizabeth Henderson. 1959; rpt. Cambridge, Mass.: Harvard University Press.

TELSER, L. G. 1972. *Competition, collusion and game theory*. Chicago: Aldine Atherton.

TRIFFIN, R. 1940. *Monopolistic competition and general equilibrium theory*. Cambridge, Mass.: Harvard University Press.

TVERSKY, A., AND D. KAHNEMAN. 1971. Belief in the law of small numbers. *Psychological Bulletin* 76:105–110.

VEBLEN, T. 1967. *The theory of the leisure class*. 1899; rpt. New York: Viking.

VEINOTT, P. 1966. The status of mathematical inventory theory. *Management Science* 12:745–777.

VICKERS, D. 1968. *The theory of the firm: production, capital and finance*. New York: McGraw-Hill.

VICKREY, W. 1961. Counterspeculation, auctions and competitive sealed tenders. *Journal of Finance* 16:8–37.

VON NEUMANN, J., AND O. MORGENSTERN. 1953. *Theory of games and economic behavior,* 3rd ed. 1944; rpt. Princeton: Princeton University Press.

WAGNER, H. M. 1969. *Principles of operations research.* Englewood Cliffs, N.J.: Prentice-Hall.

WALD, A. 1951. On some systems of equations of mathematical economics (translation). *Econometrica* 19:368–403.

WALRAS, L. 1954. *Elements of pure economics,* translated by W. Jaffe. 1874; rpt. New York: Allen and Unwin.

WHITIN, T. M. 1953. *The theory of inventory management.* Princeton: Princeton University Press.

WICKHAM, S. 1966. *Concentration et Dimensions.* Paris: Flammarion.

WILLIAMSON, O. E. 1965. Selling expense as a barrier to entry. *Quarterly Journal of Economics* 79:112–128.

———. 1975. *Markets and hierarchies: analysis and antitrust implications.* Glencoe, Ill.: Free Press.

WILSON, R. B. 1967. Competition bidding with asymmetric information. *Management Science* 13:816–820.

ZABEL, E. 1970. Monopoly and uncertainty. *Review of Economic Studies* 37:205–220.

ZEUTHEN, F. 1930. *Problems of monopoly and economic welfare.* London: Routledge.

Index

Bei Fragen zur Produktsicherheit wenden Sie sich bitte an:
If you have any questions regarding product safety,
please contact:

Walter de Gruyter GmbH
Genthiner Straße 13
10785 Berlin
productsafety@degruyterbrill.com